THE FRONTIERS COLLECTION

THE FRONTIERS COLLECTION

The books in this collection are devoted to challenging and open problems at the forefront of modern science and scholarship, including related philosophical debates. In contrast to typical research monographs, however, they strive to present their topics in a manner accessible also to scientifically literate non-specialists wishing to gain insight into the deeper implications and fascinating questions involved. Taken as a whole, the series reflects the need for a fundamental and interdisciplinary approach to modern science and research. Furthermore, it is intended to encourage active academics in all fields to ponder over important and perhaps controversial issues beyond their own speciality. Extending from quantum physics and relativity to entropy, consciousness, language and complex systems—the Frontiers Collection will inspire readers to push back the frontiers of their own knowledge.

More information about this series at http://www.springer.com/series/5342

For a full list of published titles, please see back of book or springer.com/series/5342

Juan Manuel Durán

Computer Simulations in Science and Engineering

Concepts–Practices–Perspectives

 Springer

Juan Manuel Durán
Faculty of Technology, Policy and
    Management
Delft University of Technology
Delft, The Netherlands

ISSN 1612-3018   ISSN 2197-6619   (electronic)
THE FRONTIERS COLLECTION
ISBN 978-3-030-08122-5   ISBN 978-3-319-90882-3   (eBook)
https://doi.org/10.1007/978-3-319-90882-3

This Springer imprint is published by the registered company Springer Nature Switzerland AG
The registered company address is: Gewerbestrasse 11, 6330 Cham, Switzerland

To mom, dad, and Jo
For their unconditional support and love.

To my Bee
I could have not made this journey without
you.

To Mauri
In true friendship.

To Manuel
With love and admiration.

Preface

The ubiquitous presence of computer simulations in all kinds of research areas evidences their role as the new driving force for the advancement of science and engineering research. Nothing seems to escape the image of success that computer simulations project onto the research community and the general public. One simple way to illustrate this consists in asking ourselves how would contemporary science and engineering look without the use of computer simulations. The answer would certainly diverge from the current image we have of scientific and engineering research.

As much as computer simulations are successful, they are also methods that fail in their purpose of inquiring about the world, and as much as researchers make use of them, computer simulations raise important questions that are at the heart of contemporary science and engineering practice. In this respect, computer simulations make a fantastic subject of research for the natural sciences, the social sciences, engineering and, as in our case, also for philosophy. Studies on computer simulations touch upon many different facets of scientific and engineering research and evoke philosophically inclined questions of interpretation with close ties to problems in experimental settings and engineering applications.

This book will introduce the reader, in an accessible and self-contained manner, to these various fascinating aspects of computer simulations. An historical study on the conceptualization of computer simulations throughout the past sixty years opens up the vast world of computer simulations and their implications. The focus then is shifted to the discussion on their methodology, their epistemology, and the possibilities of an ethical framework, among other issues.

The scope of this book is relatively broad in order to familiarize the reader with the many facets of computer simulations. Throughout the book, I have sought to maintain a healthy balance between the conceptual ideas associated with the philosophy of computer simulations on the one hand, and their practice in science and engineering on the other hand. To this end, this book has been conceived for a broad audience, from scientists and engineers, policy makers and academics, to the general public. It welcomes anyone interested in philosophical questions—and conceivable answers—to issues raised by the theory and practice of computer

simulations. It must be mentioned that although the book is written in a philosophical tone, it does not engage in deep philosophical discussions. Rather, it seeks
to explore the synergy between technical aspects of computer simulations and the
philosophical value there emerging. In this respect, the ideal readers of this book are
researchers across disciplines working on computer simulations but holding
philosophical inclinations. This is, of course, not to say that professional philosophers would not find in its pages problems and questions for their own research.

One beautiful thing about computer simulations is that they offer a fertile field of
research, both for researchers using the simulations as well as those reflecting upon
them. In this respect, although the book might have some merits, it also falls short
in many respects. For instance, it does not address the work of computer simulations in the social sciences, a very fruitful area of research. It also does not discuss
the use of computer simulations in and for policy making, their uses for reporting to
the general public, nor their role in a democratic society where science and engineering practice is a common good. This is certainly unfortunate. But there are two
reasons that, I hope, excuse these shortcomings. One is that I am not a specialist in
any of these fields of research, and therefore, my contribution would have been of
little interest. Each of the fields mentioned brings about specific issues in their own
right that those involved in their study know best. The second reason stems from
the fact that, as all researchers know, time and, also in this case, space are tyrant. It
would be an impossible task to even scratch the surface of the many areas where
computer simulations are active and thriving.

As a general rule for the book, I present a given topic and discuss problems and
potential solutions to it. No topic should be addressed as unrelated to any other
topic in the book, nor should a proposed answer be taken as final. In this sense, the
book aims at motivating further discussions, rather than providing a closed set of
topics and the answers to their core issues. Each chapter should nevertheless present
a self-contained discussion of a general theme of computer simulations. I must also
mention that each chapter contains profuse references to the specialized literature,
giving the reader the opportunity to pursue further his or her own interests on a
given subject.

The book is organized as follows. In Chap. 1, I address the question 'what are
computer simulations?' by giving an historical overview of the concept. Tracking
back the concept of computer simulation to the early 1960s, we will soon realize
that many contemporary definitions owe much to these early attempts. A proper
grasp of the history of the concept will turn out to be very important for the
development of a solid understanding of computer simulations. In particular, I
identify two traditions, one that puts the emphasis on implementing mathematical
models on the computer, and another for which the prominent feature is the rep-
resentationalcapacity of the computer simulation. Depending on which tradition
researchers choose to follow, the assumptions and implications to be drawn from
computer simulations will differ. The chapter ends with a discussion on the now
standard clasification of computer simulations.

The core of Chap. 2 is to introduce and discuss in detail the constituents of *simulation models*—that is, the models at the basis of computer simulations. To this end, I discuss diverse approaches to scientific and engineering models with the purpose of entrenching simulation models as a rather different kind. Once this is accomplished, the chapter goes on presenting and discussing three units of analysis constitutive of computer simulations, namely the specification, the algorithm, and the computer process. This chapter is the most technical of the book, as it draws extensively from studies on software engineering and computer science. In order to balance this with some philosophy, it also presents several problems related to these units of analysis—both individually and in relation to each other.

The sole purpose of Chap. 3 is to present the discussion on whether computer simulations are epistemologically equivalent to laboratory experimentation. The importance of establishing such equivalence has its roots in a tradition that takes experimentation as the solid foundation for our insight into the world. Since much of the work demanded of computer simulations is to provide knowledge and understanding of real-world phenomena that would otherwise not be possible, the question of their epistemological power in comparison with laboratory experimentation naturally occurs. Following the philosophical tradition of discussing these issues, I focus on the now time-honored problem of the 'materiality' of computer simulations.

Although Chaps. 4 and 5 are independent of each other, they do share the interest of establishing the epistemological power of computer simulations. While Chap. 4 does so by discussing the many ways in which computer simulations are reliable, Chap. 5 does it by showing the many epistemic functions attached to computer simulations. These two chapters, then, represent my contribution to the many attempts to ground the epistemic power of computer simulations. Let us note that these chapters are, at their basis, an answer to Chap. 2 which discusses computer simulations *vis à vis* laboratory experimentation.

Next, Chap. 6 addresses issues that are arguably less visible in the literature on computer simulations. The core question here is whether computer simulations should be understood as a third paradigm of scientific and engineering research—theory, experimentation, and Big Data being the first, second, and fourth paradigm respectively. To this end, I first discuss the use of Big Data in scientific and engineering practice, and what it means to be a paradigm. With these elements in mind, I begin a discussion on the possibilities of holding causal relations in Big Data science as well as computer simulations, and what this means for the establishment of these methodologies as paradigms of research. I finish the chapter with a comparison between computer simulations and Big Data with a special emphasis on what sets them apart.

The last chapter of the book, Chap. 7, addresses an issue that has been virtually unexplored in the literature on ethics of technology, that is, the prospect of an ethics exclusively for computer simulations. Admittedly, the literature on computer simulations is more interested in their methodology and epistemology and much less on the ethical implications that come with designing, implementing, and using computer simulations. In response to this lack of attention, I approach this chapter as an overview of the ethical problems addressed in the specialized literature.

Stuttgart, Germany   Juan Manuel Durán
June 2018

Acknowledgements

As is usually the case, there are many people who have contributed in making this book possible. First and foremost, I would like to thank Marisa Velasco, Pío García, and Paul Humphreys for their initial encouragement to write this book. All three have had a strong presence in my education as a philosopher, and this book certainly owns them much. To all three, my gratitude.

This book started in Argentina and ended in Germany. As a postdoc at the Centro de Investigaciones de la Facultad de Filosofía y Humanidades (CIFFyH), Universidad Nacional de Córdoba (UNC—Argentina), funded by the National Scientific and Technical Research Council (CONICET), I had the chance to write and discuss the first chapters with my research group. For this, I am grateful to Víctor Rodríguez, José Ahumada, Julián Reynoso, Maximiliano Bozzoli, Penelope Lodeyro, Xavier Huvelle, Javier Blanco, and María Silvia Polzella. Andrés Ilčić is another member of this group, but he deserves special recognition. Andrés read every chapter of the book, made thoughtful comments, and amended several mistakes that had gone unnoticed by me. He has also diligently checked many of the formulas I use in the book. For this and for the innumerable discussions we have had, thank you Andy. Naturally, all mistakes are my responsibility. A very sincere thank goes to CIFFyH, the UNC, and CONICET, for their support to the humanities in general and me in particular.

I also have many people to thank that, although they did not contribute directly to the book they showed their support and encouragement throughout good and bad days. My eternal gratitude goes to my good friend Mauricio Zalazar and my sister Jo. Thank you guys for being there when I need you the most. My parents have also been a constant source of support and love, thank you mom and dad, this book would not exist without you. Thanks also go to Víctor Scarafía and my newly adopted family: the Pomper and the Ebers. Thank you all guys for being so great. Special thanks goto the two Omas: Oma Pomper and Oma Eber. I love you ladies. Finally, thanks go to PeterOstritsch for his support in many aspects of my life, most of them unrelated to this book.

The book ended in the Department of Philosophy of Science and Technology of Computer Simulation, at the High-Performance Computing Center Stuttgart (HLRS), University of Stuttgart. The department was created by Michael Resch and Andreas Kaminski, and funded by the Ministerium for Wissenschaft, Forschung and KunstBaden-Württemberg (MWK), whom I thank for providing a comfortable atmosphere for working on the book. I extend my thanks to all the members of the department, Nico Formanek, Michael Hermann, Alena Wackerbarth, and Hildrun Lampe, I will never forget the many fundamental philosophical discussions we had at lunch—and over our new espresso machine—on the most varied of topics. I feel especially lucky to share an office with Nico and Michael, good friends and great philosophers. Thank you guys for checking the formulas that I included in the book. The mistakes are, again, entirely my responsibility. Thanks also go to Björn Schembera, a true philosopher in disguise, for our time discussing so many technical issues about computer simulations, some of which found a place in this book. From the visualization department at the HLRS, thanks go to Martin Aumüller, Thomas Obst, Wolfgang Schotte, and Uwe Woessner who patiently explained to me the many details of their work as well as provided the images for augmented reality and virtual reality discussed in the chapter on visualization. The images of the tornado that are also in that chapter were provided by the National Center for Supercomputing Applications, University of Illinois at Urbana-Champaign. For this, I am in great debt to Barbara Jewett for her patience, time, and great help in finding the images.

I would also like to express my gratitude to my former Ph.D. advisor, Ulla Pompe-Alama for encouragement and suggestions on early drafts. Special thanks go to Raphael van Riel and the University of Duisburg-Essen for their support on a short-term fellowship. For several reasons, directly and indirectly related to the book, I am in debt with Mauricio Villaseñor, Jordi Valverdú, Leandro Giri, Verónica Pedersen, Manuel Barrantes, Itatí Branca, Ramón Alvarado, Johannes Lenhard, and Claus Beisbart. Thank you all for your comments, suggestions, encouragement during the different stages of the book, and for having all sorts of philosophical conversations with me. Angela Lahee, my editor at Springer, deserves much credit for her patience, encouragement, and helpful support in the production of this book. Although I would have kept on polishing the ideas in this book and, equally important, my English, the time to put an end to it is upon me.

My enormous gratitude goes to Tuncer Ören, with whom I shared many correspondences on ethical and moral problems of computer simulations, leading to the final chapter of the book. Professor Ören's love and dedication to philosophical studies on computer simulation is a source of inspiration.

Finally, in times where science and technology are undoubtedly a fundamental tool for the progress of society, it is heartbreaking to see how the current government of Argentina—and in many other places in Latin-America as well—are cutting funding in science and technology research, humanities, and the social sciences. I observe with equal horror the political decisions explicitly targeted to the destruction of the educational system. I then dedicate this book to the Argentinean scientific and technological community, for they have shown time and time again their greatness and brilliance despite unfavorable conditions.

This book owes Kassandra too much. She left her impression when correcting my English, when suggesting me to re-write a whole paragraph, and when she let go a planned appointment that I forgot while finishing a section. For this, and for thousands of other reasons, this book is entirely dedicated to her.

Contents

1 **The Universe of Computer Simulations** 1
  1.1 What Are Computer Simulations? 2
      1.1.1 Computer Simulations as Problem-Solving
            Techniques 9
      1.1.2 Computer Simulations as Description of Patterns
            of Behavior................................ 13
  1.2 Kinds of Computer Simulations...................... 18
      1.2.1 Cellular Automata 20
      1.2.2 Agent-Based Simulations 23
      1.2.3 Equation-Based Simulations 25
  1.3 Concluding Remarks................................ 29

2 **Units of Analysis I: Models and Computer Simulations** 31
  2.1 Scientific and Engineering Models...................... 32
  2.2 Computer Simulations............................... 37
      2.2.1 Constituents of Computer Simulations 37
  2.3 Concluding Remarks................................ 60

3 **Units of Analysis II: Laboratory Experimentation
  and Computer Simulations** 61
  3.1 Laboratory Experimentation and Computer Simulations 62
  3.2 The Materiality Argument............................ 64
      3.2.1 The Identity of the Algorithm..................... 66
      3.2.2 Material Stuff as Criterion 69
      3.2.3 Models as (Total) Mediators...................... 75
  3.3 Concluding Remarks................................ 77

4   Trusting Computer Simulations 79
   4.1   Knowledge and Understanding 81
   4.2   Building Trust 86
      4.2.1   Accuracy, Precision, and Calibration................. 86
      4.2.2   Verification and Validation....................... 91
   4.3   Errors and Opacity 96
      4.3.1   Errors 97
      4.3.2   Epistemic Opacity 103
   4.4   Concluding Remarks................................ 110

5   Epistemic Functions of Computer Simulations 113
   5.1   Linguistic Forms of Understanding 113
      5.1.1   Explanatory Force 113
      5.1.2   Predictive Tools.............................. 121
      5.1.3   Exploratory Strategies 127
   5.2   Non-linguistic Forms of Understanding 134
      5.2.1   Visualization 134
   5.3   Concluding Remarks................................ 144

6   Technological Paradigms 147
   6.1   The New Paradigms 149
   6.2   Big Data: How to Do Science with Large Amounts of Data? 153
      6.2.1   An Example of Big Data 158
   6.3   The Fight for Causality: Big Data and Computer Simulations 161
   6.4   Concluding Remarks................................ 169

7   Ethics and Computer Simulations.......................... 171
   7.1   Computer Ethics, Ethics in Engineering, and Ethics in Science ... 172
   7.2   An Overview of the Ethics in Computer Simulations 175
      7.2.1   Williamson 175
      7.2.2   Brey 179
      7.2.3   Ören 181
   7.3   Professional Practice and a Code of Ethics................. 182
      7.3.1   A Code of Ethics for Researchers in Computer
           Simulations 184
      7.3.2   Professional Responsibilities....................... 186
   7.4   Concluding Remarks................................ 188

References 189

Introduction

In 2009, a debate erupted around the question of whether computer simulations introduce novel *philosophical* problems or if they are merely a *scientific* novelty. Roman Frigg and Julian Reiss, two prominent philosophers that ignited the debate, noted that philosophers have largely assumed some form of philosophical novelty of computer simulations without actually engaging the question of its possibility. Such an assumption rested on one simple confusion: Philosophers were thinking that scientific novelty licenses philosophical novelty. This gave course to issuing a warning over the growth of overemphasized and generally unwarranted claims about the philosophical importance of computer simulations. This growth, according to the authors, was reflected in the increasing number of philosophers convinced that the philosophy of science, nourished by computer simulations, required an entirely new epistemology, a revised ontology, and novel semantics.

It is important to point out that Frigg and Reiss are not objecting to the novelty of computer simulations in scientific and engineering practice, nor their importance in the advancement of science, but rather that simulations raise few, if any, new philosophical question. In their own words, '[t]he philosophical problems that do come up in connection with simulations are not specific to simulations and most of them are variants of problems that have been discussed in other contexts before. This is not to say that simulations do not raise new problems of their own. These specific problems are, however, mostly of a mathematical or psychological, not philosophical nature' (Frigg and Reiss 2009, 595).

I share Frigg and Reiss' puzzlement on this issue. It is hard to believe that a *new* scientific method—instrument, mechanism, etc.—however powerful as it might be, could all by itself imperil current philosophy of science and technology to the point that they need to be rewritten. But this is only true if we accept the claim that computer simulations come to *rewrite* long-standing disciplines, which I do not think is the case. To me, if we are able to reconstruct and give new meaning to old philosophical problems in light of computer simulations, then we are basically establishing their philosophical *novelty*.

Let us now ask the question in what sense are computer simulations a philosophical novelty? There are two ways to unpack the problem. Either computer simulations pose a series of philosophical questions that escape standard philosophical treatment, in which case they can be added to our philosophical corpus; or they challenge established philosophical ideas, in which case the current corpus expands standard debates into new domains. The first case has been proposed by (Humphreys 2009), whereas the second case has been argued by myself (Durán, under review). Let me now briefly discuss why computer simulations represent, in many respects, a scientific and philosophical novelty.

The core of Humphreys' argument is to recognize that we could either understand computer simulations by focusing on how traditional philosophy illuminates their study (e.g., through a philosophy of models, or a philosophy of experiment), or by focusing exclusively on aspects about computer simulations that constitute, in and by themselves, genuine philosophical challenges. It is this second way of looking at the questions about their novelty that grants philosophical importance to computer simulations.

The chief claim here is that computer simulations can solve otherwise intractable models and thus amplify our cognitive abilities. But such amplification comes with a price 'for an increasing number of fields in science, an exclusively anthropocentric epistemology is no longer appropriate because there now exist superior, non-human, epistemic authorities' (Humphreys 2009, 617). Humphreys calls this the *anthropocentric predicament* as a way to illustrate current trends in science and engineering where computer simulations are moving humans away from the center of production of knowledge. According to him, a brief overview on the history of philosophy of science shows that humans have always been at the center of production of knowledge. This conclusion includes the period of the logical and empirical positivism, where the human senses were the ultimate authority (Humphreys 2009, 616). A similar conclusion follows from the analysis of alternatives to the empiricist, such as Quine's and Kuhn's epistemologies.

When confronted with claims about the philosophical novelty of computer simulations, Humphreys points out that the standard empiricist viewpoint has prevented a complete separation between humans and their capacity to evaluate and produce scientific knowledge. The anthropocentric predicament, then, comes to highlight precisely this separation: It is the claim that humans have lost their privileged position as the ultimate epistemic authority.[1] The claim finally gets its support from the view that scientific practice only progresses because new methods are available for handling large amounts of information. Handling information, according to Humphreys, is the key for the progress of science today, which can only be attainable if humans are removed from the center of the epistemic activity (Humphreys 2004, 8).

---

[1] Humphreys makes a further distinction between scientific practice completely carried out by computers—one that he calls *the automated scenario*—and one in which computers only partially fulfill scientific activity—that is, the *hybrid scenario*. He restricts his analysis, however, to the hybrid scenario (Humphreys 2009, 616–617).

The anthropocentric predicament, as philosophically relevant as it is in itself, also brings about four extra novelties unanalyzed by the traditional philosophy of science. Those are *epistemic opacity, the temporal dynamics of simulations, semantics*, and the *in practice/in principle* distinction. All four are novel philosophical issues brought up by computer simulations; all four have no answer in traditional philosophical accounts of models and experimentation; and all four represent a challenge for the philosophy of science.

The first novelty is *epistemic opacity*, a topic that is currently attracting much attention from philosophers. Although I discuss this issue in some detail in Sect. 4. 3.2, briefly mentioning the basic assumptions behind epistemic opacity will shed some light on the novelty of computer simulations. Epistemic opacity, then, is the philosophical position that takes that it is impossible for any human to know all the epistemically relevant elements of a computer simulation. Humphreys presents this point in the following way: "A process is essentially epistemically opaque to [a cognitive agent] X if and only if it is impossible, given the nature of X, for X to know all of the epistemically relevant elements of the process" (Humphreys 2009, 618). To put the same idea in a different form, if a cognitive agent could stop the computer simulation and take a look inside, she would not be able to know the previous states of the process, reconstruct the simulation up to the point of stop, or predict future states given previous states. Being epistemically opaque means that, due to the complexity and speed of the computational process, no cognitive agent could know what makes a simulation an epistemically relevant process.

A second novelty that is related to epistemic opacity is the 'temporal dynamics' of computer simulations. This concept has two possible interpretations. Either it refers to the necessary computer time to solve the simulation model, or it stands for the temporal development of the target system as represented in the simulation model. A good example that merges these two ideas is a simulation of the atmosphere: The simulation model represents the dynamics of the atmosphere, for a year and it takes, say, ten days to compute.

These two novelties nicely illustrate what is typical of computer simulations, namely the inherent complexity of simulations in themselves, as is the case of epistemic opacity and the first interpretation of temporal dynamics; and the inherent complexity of the target systems that computer simulations usually represent, as is the case of the second interpretation of temporal dynamics. What is common between these two novelties is that they both entrench computers as the epistemic authority since they are able to produce reliable results that no human or group of humans could produce by themselves. Either because the process of computing is too complex to follow or because the target system is too complex to comprehend, computers become the exclusive source for obtaining information about the world.

The second interpretation of temporal dynamics is tailored to the novelty of the *semantics*, which asks the question of how theories and models represent the world, now adjusting the picture to fit a computer algorithm. Thus, the chief issue here is how the syntax of a computer algorithm maps onto the world, and how a given theory is actually brought into contact with data.

Finally, the distinction in *principle/in practice* is intended to sort out what is applicable in practice and what is applicable only in principle. To Humphreys, it is a philosophical fantasy to say that, in principle, all mathematical models find a solution within computer simulations (Humphreys 2009, 623). It is a fantasy because it is clearly false, although philosophers have claimed its possibility— hence, in principle. Humphreys suggests, instead, that in approaching computers, philosophers must keep a more down-to-earth attitude, limited to the technical and empirical constraints that simulations can offer.

My position is complementary to Humphreys' in the sense that it shows how computer simulations challenge established ideas in the philosophy of science. To this end, I begin by arguing for a specific way of understanding simulation models, the kind of model at the basis of computer simulations. To me, a simulation model recasts a multiplicity of models into one 'super-model.' That is to say, simulation models are an amalgam of different sorts of computer models, all having their own scales, input parameters, and protocols. In this context, I claim for three novelties in philosophy, namely *representation, abstraction, and explanation.*

About the first novelty, I claim that the multiplicity of models implies that *representation* of a target system is more holistic in the sense that it encompasses all and every model implemented in the simulation model. To put the same idea in a rather different form, the representation of the simulation model is not given by any individual implemented model but rather by the combination of all of them.

The challenge that computer simulations bring to the notion of *abstraction* and idealization is that, typically, the latter presupposes some form of *neglecting* stance. Thus, *abstraction* aims at ignoring concrete features that the target system possesses in order to focus on their formal setup; *idealizations*, on the other hand, come in two flavors: While Aristotelian idealizations consist in 'stripping away' properties that we believe not to be relevant for our purposes, Galilean idealizations involve deliberate distortions. Now, in order to implement the required variety of models into a single simulation model, it is important to count on techniques by which information is hidden from the users, but not neglected from the models (Colburn and Shute 2007). This is to say that the properties, structures, operations, relations, and the like present in each mathematical model can be effectively implemented into the simulation model without stating explicitly how such implementation is carried out.

Finally, *scientific explanation* is a time-honored philosophical topic where much has been said. When it comes to explanation in computer simulations, however, I propose a rather different look at the issue than the standard treatment offers. One interesting point here is that, in the classic idea that explanation is of a real-world phenomenon I oppose the claim that explanation is, first and foremost, of the results of computer simulations. In this context, many new questions emerge seeking an answer. I discuss scientific explanation in more detail in Sect. 5.1.1.

As I have mentioned before, I do believe that computer simulations raise novel questions for the philosophy of science. This book is living proof of that belief. But even if we do not believe in their philosophical novelty, we still need to understand computer simulations as scientific novelties with a critical and philosophical eye. To these ends, this book presents and discusses several theoretical and philosophical issues at the heart of computer simulations. Having said all of this, we may now submerge ourselves into their pages.

Chapter 1
The Universe of Computer Simulations

The universe of computer simulations is vast, flourishing in almost every scientific discipline, and still resisting a general conceptualization. From the early computations of the Moon's orbit carried out by punched card machines, to the most recent attempts to simulate quantum states, computer simulations have a uniquely short but very rich history.

We can situate the first use of a machine for scientific purposes in England at the end of the 1920s. More precisely, it was in 1928 when the young astronomer and pioneer in the use of machines Leslie J. Comrie predicted the motion of the Moon for the years 1935–2000. During that year, Comrie made intensive use of a Herman Hollerith punched card machine to compute the summation of harmonic terms in predicting the Moon's orbit. Such groundbreaking work would not stay in the shadows, and by the mid 1930s it had cross the ocean to Columbia University in New York City. It was there that Wallace Eckert founded a laboratory that made use of punched card tabulating machines—now built by IBM—to perform calculations related to astronomical research, including of course an extensive study of the motion of the Moon.

Both Comrie's and Eckert's uses of punched card machines share a few commonalities with today's use of simulations. Most prominently, both implement a special kind of model that describes the behavior of a target system, and which can be interpreted and computed by a machine. While Comrie's computing rendered data about the motions of the Moon, Eckert's simulation described planetary movement.

These methods certainly pioneered and revolutionized their respective fields, as well as many other branches of the natural and social sciences. However, Comrie's and Eckert's simulations significantly differ from today's *computer simulations*. Upon closer inspection, differences can be found everywhere. The introduction of silicon based circuits, as well as the subsequent standardization of the circuit board, made a significant contribution to the growth of computational power. The increase in the speed of calculation, size of memory, and expressive power of programming language forcefully challenged the established ideas on the nature of computation

© Springer Nature Switzerland AG 2018
J. M. Durán, *Computer Simulations in Science and Engineering*,
The Frontiers Collection, https://doi.org/10.1007/978-3-319-90882-3_1

and of its domain of application. Punched card machines rapidly became obsolete as they are slow in speed, unreliable in their results, limited in their programming, and based on stiff technology (e.g., there were very few exchangeable modules). In fact, a major disadvantage of the punched card over modern computers is that they are error-prone and time-consuming machines, and therefore the reliability of their results as well as their representational accuracy is difficult to ground. However, perhaps the most radical difference between Comrie's and Eckert's simulations, on the one hand, and modern computer simulations on the other, is the automation process that characterizes the latter. In todays computer simulations, researchers are losing ground on their influence and power to interfere in the process of computing, and this will become more prominent as complexity and computational power increases.

Modern computers come to amend many aspects of scientific and engineering practice with more precise computations, and more accurate representations. Accuracy, computational power, and reduction of errors are, as we will see, the main keys of computer simulations that unlock the world.

In light of contemporary computers, then, it is not correct to maintain that Comrie's prediction of the motion of the Moon and Eckert's solution of planetary equations are computer simulations. This is, of course, not to say that they are not simulations at all. But in order to accommodate to the way scientists and engineers use the term today, it is not sufficient to be able to compute a special model or to produce certain kinds of results about a target system. Speed, storage, language expressiveness, and the capacity to be (re)programmed are chief concepts for the modern notion of computer simulation.

What are computer simulations then? This is a philosophically motivated question that has found different answers from scientists, engineers, and philosophers. The heterogeneity of their answers makes explicit how differently each researcher conceives computer simulations, how their definitions vary from one generation to another, and how difficult it is to come up with a unified notion. It is important, however, to have a good sense of their nature. Let us discuss this in more extent.

1.1   What Are Computer Simulations?

Recent philosophical literature takes computer simulations as aids for overcoming imperfections and limitations of human cognition. Such imperfections and limitations are tailored to the natural human constraints of computing, processing and classifying large amounts of data. Paul Humphreys, one of the first contemporary philosophers to address computer simulations from a purely philosophical viewpoint, takes them as an 'amplification instrument,' that is, one that speeds up what the unaided human could not do by herself (Humphreys 2004, 110). In a similar sense, Margaret Morrison, yet another central figure in philosophical studies on computer simulations, considers that although they are another form of modeling, "given the various functions of simulation [...] one could certainly characterize it as a type of 'enhanced' modelling" (Morrison 2009, 47).

Both claims are fundamentally correct. Computer simulations compute, analyze, render, and visualize data in many ways that are unattainable for any group of humans. Contrast, for instance, the time required for a human to identify potential antibiotics for infections diseases such as anthrax, with a simulation of the ribosome in motion at atomic detail (Laboratory 2015). Or, if preferred, compare any set of human computational capabilities with the supercomputers used at the High Performance Computing Center Stuttgart, home of the Cray XC40 Hazel Hen with a peak performance of 7.42 Petaflops and a memory capacity of 128 GB per node.[1]

As pointed out by Humphreys and Morrison, there are different senses in which computer simulations enhance our capacities. This could be by amplifying our calculation skills, as Humphreys suggests, or it could be by enhancing our modeling abilities, as Morrison suggests.

One would be naturally inclined to think that computer simulations amplify our computational capability as well as enhance our modeling abilities. However, a quick look at the history of the concept shows otherwise. To some authors, a proper definition must highlight the importance of finding solutions to a model. To others, the right definition centers the attention to describing patterns of behavior of a target system. Under the first interpretation, the computational power of the machine allows us to solve models that, otherwise, would be analytically intractable. In that respect, a computer simulation 'amplifies' or 'enhances' our cognitive capacities by providing computational power to what is beyond our cognitive reach. The notion of computer simulation is then dependent on the physics of the computer and furnishes the idea that technological change expands the boundaries of scientific and engineering research. Such a claim is also historically grounded. From Hollerith's punched card machines to the silicon-based computer, the increment of the physical power of computers has enabled scientists and engineers to find different solutions to a variety of models. Let me call this first interpretation *the problem-solving viewpoint* on computer simulations.

Under the second interpretation, the emphasis is on the capacity of the simulation to describe a target system. For this, we have a powerful language that represents, to certain acceptable degrees of detail, several levels of description. In that respect, a computer simulation 'amplifies' or 'enhances' our modeling abilities by providing more accurate representation of a target system. Thus understood, the notion of computer simulation is tailored to the way in which they describe a target system, and thus on the computer language used, modularization methods, software engineering techniques, etc. I call this second interpretation the *description of patterns of behavior* viewpoint on computer simulations.

Because both viewpoints emphasize different—although not necessarily incompatible—interpretations of computer simulations as enhancers, some distinc-

---

[1]It is worth noting that our neuronal network activity is, in some specific cases, faster than any supercomputer. According to relatively recent publication, Japan's *Fujitsu K* computer, consisting of 82,944 processors, takes about 40 min to simulate one second of neuronal network activity in real, biological time. In order to partially simulate the human neural activity, researchers create about 1.73 billion virtual nerve cells that were connected to 10.4 trillion virtual synapses (Himeno 2013).

tions can be drawn. For starters, under the problem-solving viewpoint, computer simulations are not experiments in any traditional sense, but rather the manipulation of an abstract and formal structure (i.e., mathematical models). In fact, to many advocates of this viewpoint, experimental practice is confined to the traditional laboratory as computer simulations are more of a crunching numbers practice, closer to mathematics and logic. The description of patterns of behavior viewpoint, on the other hand, allows us to treat computer simulations as experiments in a straightforward sense. The underlying intuition is that by means of describing the behavior of a target system, researchers are capable of carrying out something very similar to traditional experimental practice, such as measuring values, observing quantities, and detecting entities.

Understanding things this way has some kinship with the methodology of computer simulations. As I discuss later, the problem-solving technique viewpoint considers a simulation to be the *direct* implementation of a model on a physical computer. That is to say, mathematical models are implemented on the computer *simpliciter*. The description of patterns of behavior viewpoint, instead, holds that computer simulations have a proper methodology which it is rather different from anything we have seen in the scientific and engineering arena. These methodological differences between the two viewpoints turn out to be central for later disputes about the novelty of computer simulations in scientific and engineering research.

Another difference between these two viewpoints lies in the reasons for using computer simulations. Whereas the problem-solving viewpoint asserts that the use of computer simulations is only pragmatically justified when the model cannot be solved by more traditional methods, the description of patterns of behavior viewpoint considers that computer simulations offer valuable insight into the target system despite its analytical intractability. Let us note that what is also at stake here is the epistemic priority of one method over another. If the use of computer simulations is only justified when the model cannot be analytically solved—as many advocates of the problem-solving viewpoint claim—then analytic methods are epistemically superior to computational. This configures a specific standpoint regarding the place that computer simulations have in the scientific and engineering agenda. In particular, it significantly downplays the reliability of computer simulations for research in uncharted territory. We will have more to say about this throughout this book.[2]

Finally, tackling problems in computer simulations can be very different depending on the viewpoint adopted. For the problem-solving viewpoint, any issues related to the results of the simulation (e.g., accuracy, computability, representability, etc.) can be solved on technical grounds (i.e., by increasing speed and memory, changing the underlying architecture, etc.). Instead, for the advocate of the description of patterns of behavior, the same issues have an entirely different treatment. Incorrect results, for instance, are approached by analyzing practical considerations at the design level, such as new specifications for the target system, alternative assessments

---

[2]Many philosophers have tried to understand the nature of computer simulations. What I have offered above is just one possible characterization. For more, the reader could refer to the following authors: Winsberg (2010, 2015), Vallverdú (2014), Morrison (2015), and Saam (2016).

of expertise knowledge, new programming languages, etc. In the same vein, incorrect results might be due to misrepresentation of the target system at the design, specification, and programming stages (see Sect. 2.2), or misrepresentations at the computational stage (e.g., errors during computing time—see Sect. 4.3). The researcher's general understanding as well as solution to these issues changes significantly depending on the viewpoint adopted.

To illustrate some of the points made so far, take a simple equation-based computer simulation of the dynamics of satellite orbiting around a planet under tidal stress. To simulate such dynamics, researchers typically begins with a mathematical model of the target system. A good model is furnished by classical Newtonian mechanics, as described by Woolfson and Pert in (1999a).

For a planet of mass M and a satellite of mass m ($\ll M$), in an orbit of semi-major axis a and eccentricity e, the total energy is

$$E = -\frac{GMm}{2a} \tag{1.1}$$

and the angular momentum is

$$H = \{GMa(1 - e^2)\}m \tag{1.2}$$

If E is to decrease, then a must become smaller; but if H is constant, then e must become smaller—that is to say, that the orbit must round off. The quantity which remains constant is $a(1 - e^2)$, the *semi-latus rectum* as shown in Fig. 1.1. The planet is described by a point mass, P, and the satellite by a distribution of three masses, each $m/3$, at positions S_1, S_2 and S_3, forming an equilateral triangle when free of stress. The masses are connected, as shown, by springs, each of unstressed length l and the same spring constant, k (Fig. 1.2). Thus a spring constantly stretched to a length l' will exert an inward force equal to

$$F = k(l' - l) \tag{1.3}$$

It is also important to introduce a dissipative element into the system by making the force dependent on the rate of expansion or contraction of the spring, giving the following force law:

$$F = k(l' - l) - c\frac{dl'}{dt} \tag{1.4}$$

where the force acts inwards at the two ends. It is the second term in Eq. 1.4 which gives the simulation of the hysteresis losses in the satellite (Woolfson and Pert 1999a, 18–19).

This is a model for a computer simulation of a satellite in orbit around a planet subject to tidal stress. The satellite stretches along the radius vector in a periodic fashion, provided that the orbit is non-circular. Given that the satellite is not perfectly

Fig. 1.1 The elliptical orbit of a satellite relative to the planet at one focus. Points q and Q are the nearest and furthest points from the planet, respectively. (Woolfson and Pert 1999a,19)

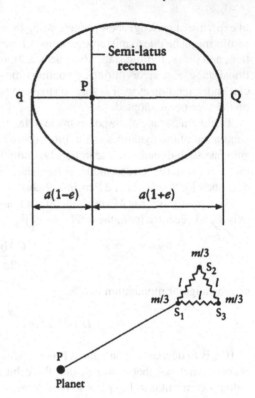

Fig. 1.2 The satellite is described by three masses, each $m/3$, connected by springs each of the same unstrained length, l. (Woolfson and Pert 1999a,19)

elastic, there will be hysteresis effects and some of the mechanical energy will be converted into heat and radiated away. For all purposes, nevertheless, the simulation fully specifies the target system.

Equations 1.1 through 1.4 are a general description of the target system. Since the intention is to simulate a specific real-world phenomenon with concrete features, it must be singled out by setting the values for the parameters of the simulation. In the case of Woolfson and Pert, they use the following set of parameter values (Woolfson and Pert 1999a, 20):

1. number of bodies = 4
2. mass of the first body (planet) = 2×10^{27} kg
3. mass of satellite = 3×10^{22} kg
4. initial time step = 10 s
5. total simulation time = 125000 s
6. body chosen as origin = 1
7. tolerance = 100 m
8. initial distance of satellite = 1×10^8 m
9. unstretched length of spring = 1×10^6 m
10. initial eccentricity = 0.6

Fig. 1.3 The orbital eccentricity as a function of time (Woolfson and Pert 1999a, 20)

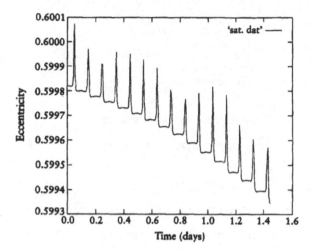

These parameters set a computer simulation of a satellite of the size of Triton, the largest moon of Neptune orbiting around a planet with a mass close to Jupiter's including, of course, a specific tidal stress, hysteresis effects, and so forth. If the parameters were changed, then naturally the simulation is of another phenomenon—though still of a two-body interaction using Newtonian mechanics (Fig. 1.3).

Here is an extract of the code corresponding to the above mathematical models as programmed in FORTRAN by Woolfson and Pert.[3]

```
PROGRAM NBODY

 [...]

  C   THE VALUES OF A AND E ARE CALCULATED EVERY 100 STEPS
  C   AND ARE STORED TOGETHER WITH THE TIME.
  C

      IST=IST+1
      IF((IST/100)*100.NE.IST)GOTO 50
      IG=IST/100
      IF(IG.GT.1000)GOTO 50


  C
  C   FIRST FIND POSITION AND VELOCITY OF CENTRE OF MASS OF THE
  C   SATELLITE.
  C
      DO 1 K=1,3
      POS(K)=0
      VEL(K)=0
      DO 2 J=2,NB
```

---

[3]For the full code, see Woolfson and Pert (1999b).

```
      POS(K)=POS(K)+X(J,K)
      VEL(K)=VEL(K)+V(J,K)
    2 CONTINUE
      POS(K)=POS(K)/(NB-1.0)
      VEL(K)=VEL(K)/(NB-1.0)
    1 CONTINUE

C
C  CALCULATE ORBITAL DISTANCE
C
      R=SQRT(POS(1)**2+POS(2)**2+POS(3)**2)

C
C  CALCULATE V**2
C
      V2=VEL(1)**2+VEL(2)**2+VEL(3)**2
C
C  CALCULATE INTRINSIC ENERGY
C
      TOTM=CM(1)+CM(2)+CM(3)+CM(4)
      EN=-G*TOTM/R+0.5*V2
```

[...]

```
      SUBROUTINE START
      DIMENSION X(20,3),V(20,3),ASTORE(1000),TSTORE(1000),
     +ESTORE(1000),XTEMP(2,20,3),VTEMP(2,20,3),CM(20),XT(20,3),
     +VT(20,3),DELV(20,3)
```

[...]

```
      READ(5,*)NORIG
```

[...]

```
C
C  INITIAL VELOCITIES ARE CALCULATED FOR THE THREE COMPONENTS OF THE
C  SATELLITE SO THAT THE SPIN ANGULAR VELOCITY OF THE SATELLITE IS
C  APPROXIMATELY EQUAL TO THE ORBITAL ANGULAR VELOCITY.
C
      VV=SQRT(G*(CM(1)+CM(2)+CM(3)+CM(4))*(1-ECC)/D)
      V(2,2)=VV*DIS2/D
      V(3,2)=VV*DIS3*COS(ANGLE)/D
```

[...]

```
RETURN
END
```

As discussed earlier, one of the characteristics of the problem-solving viewpoint is that the above mathematical model can be directly implemented on the computer. That is to say, that the code shown here corresponds in full extent to the mathematical model. In principle, nothing—relevant—is added and nothing—relevant—is eliminated. Thus, the only reason to use a computer is to find the set of solutions of the model in ways that are faster and cheaper. The description of patterns of behavior viewpoint, on the other hand, acknowledges the existence of a process for implementing the model as a computer simulation, which includes to turn the mathematical model into a rather different kind of model.

As we can see, both positions hold good grounds. The problem-solving viewpoint is correct in claiming that computer simulations must reflect the mathematical model implemented, otherwise problems of representation, reliability, and the like emerge. The description of patterns of behavior viewpoint, on the other hand, reflects scientific and engineering practice more unambiguously.

Before continuing, here is a good place to introduce some new terminology. Let us call 'mathematical models' those models used in scientific and engineering fields that make use of mathematical language. Examples of these are the above equations. Let us call 'simulation models' those models implemented on a computer—as a computer simulation—which make use of a programming language.[4] An example of a simulation model is the code shown above.

I began this chapter making a distinction between two viewpoints of computer simulations. On the one hand, the problem-solving viewpoint which emphasizes the computational side of simulations; on the other hand, the description of patterns of behavior viewpoint which places emphasis on the representations of target systems. As mentioned before, the problem-solving viewpoint does not neglect the representation of the target system, nor does the description of patterns of behavior viewpoint fail to consider computation as a core issue of computer simulations. Neither viewpoint reflect an 'all or nothing' stand—a good example of this is the definition given by Thomas H. Naylor, Donald S. Burdick, and W. Earl Sasser where they defend a patterns of behavior viewpoint on page 1361 of (Naylor et al. 1967b) and also subscribe to the problem-solving viewpoint later on page 1319. The difference between these two viewpoints lies, again, in the main features highlighted by each account. Let us see if we can make this distinction more clear.[5]

1.1.1 Computer Simulations as Problem-Solving Techniques

Under the problem-solving viewpoint, computer simulations typically exhibit some of the following features. First, simulations are adopted for cases where the target system is too complex to be analyzed on its own—call it the *complexity feature*. Second, simulations are useful for cases where the underlying mathematical model cannot be analytically solved—call it the *unanalyticity feature*. Third, mathematical models are directly implemented on the computer—call it the *direct implementation*

---

[4]Strictly speaking, a simulation model is a more complex structure consisting, among other things, of a *specification* coded in a programming language as an *algorithm* and finally implemented as a *computer process*. Although the same specification can be written in different programing languages and implemented by different computer architectures, they are all considered the same simulation model. Thus understood, the programming language by itself does not determine the notion of 'simulation model.' I shall discuss these issues in more extent in Chap. 2.

[5]An interesting introduction to the history of computer science can be found in the work of Ceruzzi (1998), De Mol et al. (2014, 2015), and particularly on computer simulations Ören (2011a, b).

feature. The complexity and unanalyticity features emphasize our human limitations for analyzing certain kind of mathematical models, at the same time enhancing the computational power simulations as a virtue. The direct implementation feature accompanies these ideas by claiming that there is no mediating methodology between the mathematical model and the physical computer. Rather, equations from the mathematical model are implemented—or solved—*simpliciter* on the physical computer in the form of a computer simulation.

The early literature of the problem-solving viewpoint presents a rather uniform perspective on the matter. For the most part, professional philosophers, scientists, and engineers see the computational power of simulations as the key unlocking their epistemic power. A good first example is the definition provided by Claude McMillan and Richard Gonzáles in 1965. In their work, the authors state four characteristic points of simulations, namely

1. Simulation is a problem solving technique.
2. It is an experimental method.
3. Application of simulation is indicated in the solution of problems of (a) systems design (b) systems analysis.
4. Simulation is resorted to when the systems under consideration cannot be analyzed using direct or formal analytical methods. (McMillan et al. 1965)

This definition is, to the extent that I could find, the first that openly conceives computer simulations as problem-solving techniques. This is not only because the authors explicitly state so in their first point, but because the definition adopts two of the three standard feature of this viewpoint. Point 4 is explicit about the use of simulations for finding solutions to otherwise unsolvable mathematical models, while point 3 suggests the adoption of simulations for system design and system analysis as they are too complex to be analyzed by their own (i.e., the complexity feature).

A year later, Daniel Teichroew and John Francis Lubin presented their own definition. Interestingly, this definition makes three features of this viewpoint more visible than any other definition in the literature. The authors begin by identifying what they call 'simulation problems,' that is, problems that are treated by simulation techniques—we shall discuss next what these techniques are. A simulation problem is basically a mathematical problem with many variables, parameters, and functions that cannot be treated analytically (i.e., the complexity feature) and thus computer simulations are the only available resource to researchers (i.e., the unanalyticity feature). The third feature, direct implementation of a mathematical model, can be found in several places in the article. In fact, the authors categorize two kinds of models, namely, *continuous change models* (i.e., those making use of Partial Differential Equations or Ordinary Differential Equations) and *discrete change models* (i.e., those models where changes in the state of the system are discrete) (Teichroew and Lubin 1966, 724). To the authors, both kinds of models are implemented directly as a computer simulation. In the authors' own words,

> Simulation problems are characterized by being mathematically intractable and having resisted solution by analytic methods. The problems usually involve many variables, many

parameters, functions which are not well-behaved mathematically, and random variables. Thus, simulation is a technique of last resort. Yet, much effort is now devoted to 'computer simulation' because it is a technique that gives answers in spite of its difficulties, costs and time required. (Teichroew and Lubin 1966, 724)

There is a further interesting claim to highlight here. Let us notice that the authors make plain the sentiment that advocates of this viewpoint have with regards to computer simulations: they are a technique of last resort.[6] That is to say, the use of computer simulations is only justified when analytic methods are unavailable. But this is more an epistemological prejudice against computer simulations than an established truth. Recent work done by philosophers shows that, in many instances, researchers prefer computer simulations over analytic methods. This, of course, for the obvious cases when the target system is intractable—as Teichroew and Lubin correctly point out—and where analytic solutions are not available. Vincent Ardourel and Julie Jebeile argue that computer simulations might even be superior to analytical solutions for the purpose of making quantitative predictions. According to these authors, "some analytical solutions make numerical applications difficult or impossible (...) analytical solutions are sometimes too sophisticated with respect to the problem at stake (...) [and] analytical methods do not offer a generic approach for solving equations like [computer simulations do]".[7] (Ardourel and Jebeile 2017, 203).

Now, advocates of the problem-solving viewpoint are also present in contemporary literature. A largely objected definition— which, despite the author's change of mind somehow managed to become a standard in the literature—is Humphreys' *working definition*: "A computer simulation is any computer-implemented method for exploring the properties of mathematical models where analytic methods are unavailable" (Humphreys 1990, 501).

Here, Humphreys gives us two features of computer simulations as problem-solvers. These are, simulations are mathematical models implemented on a computer, and they are used when analytic methods are unavailable. So far, Humphreys is a classic advocate of the problem-solving technique. A closer look of the definition, however, shows Humphreys' worries also include the nature of computation. Here is why.

Earlier, I mentioned that Naylor, Burdick, and Sasser stated that computer simulations are *numerical methods* implemented on the computer. Humphreys, instead, conceives computer simulations as *computer-implemented* methods. The distinction is not otiose, since it says something about the nature of computing a model. In fact, Humphreys urges to keep three different notions separate: *numerical mathematics*, *numerical methods*, and *numerical analysis*. Numerical mathematics is the branch of mathematics concerned with obtaining numerical values of the solutions to a given mathematical problem. Numerical methods, on the other hand, are numerical

---

[6]Regarding this last point, Prof. Ören has organized in 1982 a NATO Advanced Study Institute in Ottawa focused on addressing the context for the uses of computer simulations (personal communication). See for instance, the articles published in (Ören et al. 1982; Ören 1984).

[7]The authors identify 'numerical methods' with 'computer simulations' (Ardourel and Jebeile 2017, 202). As I show next, these two concepts must remain separate. However, this does not represent an objection to their main claim.

mathematics concerned with finding an approximate solution to the model. Finally, numerical analysis is the theoretical analysis of numerical methods and the computed solutions (Humphreys 1990, 502). Numerical methods, by themselves, cannot directly be related to computer simulations. At least two additional features must be included. First, numerical methods must be applied to a specific scientific problem. This is important because the model implemented is not *any* model, but of a specific kind (i.e., scientific and engineering models). In this way, there is no room for conflating computer simulations carried out in a scientific facility with computer simulation carried out for artistic purposes. Second, the method must be implemented on a real computer as well as computable in real time. This second feature ensures that the model is suitable for computation and complies with minimal standards of scientific research (e.g., that the computation ends within a reasonable time-frame, that the results are accurate within a certain range, etc.)

Despite being suggested only as a working definition, Humphreys received fierce objections that virtually forced him to change his original position. A chief critic was Stephan Hartmann, who objected that Humphreys' definition missed the dynamic nature of computer simulations. Hartmann, then, offered his own definition:

> Simulations are closely related to dynamic models. More concretely, a simulation results when the equations of the underlying dynamic model are solved. This model is designed to imitate the time-evolution of a real system. To put it another way, *a simulation imitates one process by another process*. In this definition, the term "process" refers solely to some object or system whose state changes in time. If the simulation is run on a computer, it is called a *computer simulation* (Hartmann 1996, 83—emphasis in original).

Simplifying this definition, we could say that a computer simulation consists of finding the set of solutions to a dynamic model by using a physical computer. Let us highlight a few interesting assumptions. First, the dynamic model is conceived as holding no differences from a mathematical model. Thus understood, the model used by M. M. Wolfson and G. J. Pert for simulating the dynamics of a satellite orbiting around a planet under tidal stress *is* the model implemented on the physical computer. Second, Hartmann is not too worried about which methods are used for solving the dynamic model. Paper and pencil, numerical methods, and computer-implemented methods seem to be all equally suitable. This concern stems from taking that the same dynamic model is solved by a human agent as well as the computer. Similarly to Naylor, Burdick, and Sasser, such an assumption raises questions about the nature of computation.

It is interesting to note that Hartmann's definition has been warmly welcomed by the philosophical community. The same year, Jerry Banks, John Carson, and Barry Nelson presented a definition similar to Hartmann's, also emphasizing the idea of *the dynamics of a process over time*, and of representation as *imitation*. They define it in the following way "[a] simulation is the imitation of the operation of a real-world process or system over time. Whether done by hand or on a computer, simulation involves the generation of an artificial history of a system and the observation of that artificial history to draw inferences concerning the operating characteristics of the real system" (Banks et al. 2010, 3). Francesco Guala also follows Hartmann in distinguishing between *static* and *dynamic* models, time-evolution of a system, and the use of simulations for mathematically solving the implemented model (Guala 2002). More recently, Wendy Parker has made explicit reference to it by characterizing a simulation as "a time-ordered sequence of states that serves as a representation of some other time-ordered sequence of states" (Parker 2009, 486).

Now, despite of the differences between Humphreys and Hartmann, they agree on a few issues as well. In fact, they both consider computer simulations as high-speed calculation equipment capable of enhancing our analytical capacity to solve otherwise unsolvable mathematical models. After Hartmann's initial objections, Humphreys coined a new definition, this time based on the notion of *computational template*. I shall discuss templates in the next section, as I believe this new conceptualization of computer simulations qualifies better for the descriptions of patterns of behavior viewpoint.

An illuminating summary is found in the work of Roman Frigg and Julian Reiss. According to the authors, there are two senses in which the notion of *computer simulation* is defined in current literature. There is a *narrow sense*, where "'simulation' refers to the use of a computer to solve an equation that we cannot solve analytically, or more generally to explore mathematical properties of equations where analytical methods fail". There is also a *broad sense*, where the term "'simulation' refers to the entire process of constructing, using, and justifying a model that involves analytically intractable mathematics" (Frigg et al. 2009, 596). To me, both senses could be included as part of the problem-solving techniques viewpoint of computer simulations.

Both categories are certainly meritorious and illuminating. Both generally capture the many senses in which philosophers of the problem-solving viewpoint define the notion of *computer simulation*. While the narrow sense focuses on the heuristic capacity of computer simulations, the broad sense emphasizes the methodological, epistemological, and pragmatic aspects of computer simulations as problem-solvers. Let us now move to a different way to conceptualize computer simulations.

1.1.2   Computer Simulations as Description of Patterns of Behavior

The view of computer simulations as problem solving techniques contrasts with the view of simulations as description of patterns of behavior. Under this view, computer simulations are primarily concerned with describing the behavior of a target system to which they develop or unfold. As mentioned before, this is not to say that the computing power of simulations is downplayed in any sense. Computer simulations as problem solvers got this point right in the sense that speed, memory, and control are core factors that emphasize the novelty of simulations in scientific and engineering practice. However, under this viewpoint, the computational power of simulations is considered a second-level feature. In this sense, instead of locating the epistemological value of computer simulations in their capacity for solving a mathematical model, their value comes from describing patterns of behavior of target systems.

Now, what are patterns? I take them to be descriptions that reflect structures, attributes, performance, and the general behavior of the target system in a specific language. More specifically, these structures, attributes, and so on are interpreted as concepts used in the sciences (e.g., H_2O, mass, etc.), causal relationships (e.g., the collision of two billiard balls), natural and logical necessities (e.g., that no enriched uranium sphere has a mass greater than $100,000 \text{ kg}$[8]), laws, principles, and constants of nature. In short, patterns are descriptions of a target system which make use of the scientific and engineering vocabulary. Naturally, these patterns also rely on expert knowledge, 'tricks of the trade', past experiences, and individual, societal and institutional preferences. In this sense, for this viewpoint, computer simulations are a conglomerate of concepts, formulae, and interpretations that facilitate the description of the patterns of behavior of a target system.

The difference in conceptualizing computer simulations in this way, as opposed to the problem-solving viewpoint, is that the physical features of the computer are no longer the primary epistemic value of computer simulations. Rather, it is their capacity to describe patterns of behavior of a target system that carries the burden. Mimicking the previous section, let us begin with some early definitions.

In 1960, Martin Shubik defined a simulation in the following way:

> A simulation of a system or an organism is the operation of a model or simulator which is a representation of the system or organism. (...) The operation of the model can be studied and, from it, properties concerning the behavior of the actual system or its subsystem can be inferred. (Shubik 1960, 909)

Shubik is highlighting two main features that are at the heart of this view. That is, that a simulation is a representation or description of the behavior of a target system, and that properties of such a target system can be inferred. The first feature is central to this viewpoint, to the extent that it gives the name to it. Emphasizing the representational capacity of simulations, as opposed to mathematical models, suggests that they are somehow different. As we shall see later, this difference lies in the number of transformations under which a mathematical model—or rather, a series of mathematical models—go through in order to result in a computer simulation. The second feature, on the other hand, highlights the use of computer simulations as proxies for understanding something about the target system. This is to say that researchers are capable of inferring properties of the target system based on the results of the simulation.

Both features, we must notice, are absent in the problem-solving viewpoint. The contrary, however, is not true. As mentioned earlier, understanding computer simulations this way does not disavow some claims of the problem solving techniques

---

[8]A clarification is due here. A computer is not *technologically* impaired to simulate an enriched uranium sphere with a mass greater than $100,000 \text{ kg}$. Rather, the kind of constraints we find in computers are related to their own physical limitations and those indicated by theories of computation. Now, given that researchers want to simulate a *real* target system, they must describe it as accurately as possible. If that target system is a natural system, such as a uranium sphere, then accuracy dictates that the simulation is limited on the mass of the sphere.

viewpoint. In particular, the capacity of computing complex models is a characteristic of computer simulations usually present in all definitions. For instance, Shubik says: "The model is amenable to manipulations which would be impossible, too expensive or impracticable to perform on the entity it portrays" (Shubik 1960, 909). It is interesting to note that, as we move forward in time, concerns about the computational power tend to disappear.

Almost two decades later, in 1979, G. Birtwistle formulated the following definition for computer simulations:

> Simulation is a technique for representing a dynamic system by a model in order to gain information about the underlying system. If the behaviour of the model correctly matches the relevant behaviour characteristics of the underlying system, we may draw inferences about the system from experiments with the model and thus spare ourselves any disasters. (Birtwistle 1979, 1)

Similarly to what Teichroew and Lubin presented with the problem-solving viewpoint, Birtwistle is also making plain the chief features of the description of patterns of behavior viewpoint. From the above definition it is clear that central to computer simulations is the representation of a target system; provided the right representation, then, researchers can draw inferences about such target system. Let us also note that, unlike Teichroew and Lubin who consider computer simulations a last resource, to Birtwistle it is a crucial piece in scientific research that helps to prevent disasters. The opposite attitudes towards computer simulations cannot find two better representatives.

Another definition worth mentioning comes from Robert E. Shannon, an industrial engineer who has worked very profusely on clarifying the nature of computer simulations (see his work from Shannon 1975 and Shannon 1978).

> We will define simulation as the process of designing a model of a real system and conducting experiments with this model for the purpose of understanding the behavior of the system and/or evaluating various strategies for the operation of the system. Thus it is critical that the model be designed in such a way that the model behavior mimics the response behavior of the real system to events that take place over time. (Shannon 1998, 7)

Again we can see how Shannon highlights the importance of representing a target system, as well as the ability to infer—and evaluate—our knowledge from computer simulations. What is perhaps the most outstanding aspect of Shannon's definition is the marked emphasis on the methodology of computer simulations. To him, it is critical that the model in the simulation mimics the behavior of the target system. It is not enough, as it is found in other authors, that the model correctly describes the relevant behavior of the target system. Attention must be given to the way in which the simulation is designed, because it is there where we will find grounds—and problems—for drawing inferences about the target system.

These latter ideas continue, more or less successfully, in the subsequent literature related to this viewpoint. A good example is Paul Humphreys' 2004 book, where he presents a detailed account of the methodology of computer simulations. Eric Winsberg, a few years later, also made an interesting effort to show the ways in which design decisions affect epistemological evaluations. According to Winsberg, present and past design decisions ground our confidence in the results of computer simulations. Let us now discuss their positions in more detail.[9]

Earlier, I mentioned that in 1990 Humphreys elaborated on a *working definition* for computer simulations. In spite of having presented it only as a working definition, he received vigorous objections that virtually forced him to change his original viewpoint on computer simulations. One of the chief critics was Stephan Hartmann, who pointed out that his working definition missed the dynamic nature of computer simulations. After Hartmann's initial objections, Humphreys coined a new definition, this time based on the notion of *computational model*.

According to his new characterization, computer simulations rely on an underlying computational model that involves representations of a target system. At first glance, this definition looks very much like the standard definitions discussed so far. However, the devil is in the details. In order to fully appreciate Humphreys' turn, we must dissect his definition of *computational model*, understood as the sextuple:

> *<computational template, construction assumptions, correction set, interpretation, initial justification, output representation>*[10]

A *computational template* is, in fact, the result of a computationally tractable theoretical template. A *theoretical template*, in turn, is the kind of very general mathematical description that can be found in a scientific work. This includes partial differential equations, such as elliptic (e.g., Laplace's equation), parabolic (e.g., the diffusion equation), hyperbolic (e.g., the wave equation), and ordinary differential equations, among others. An illuminating example of a theoretical template is Newton's Second Law, for it describes a very general constraint on the relationship between forces, mass, and acceleration. The core characteristic of theoretical templates is that researchers could specify them in a number of different ways. For instance, the force function in Newton's Second Law could either be a gravitational force, an electrostatic force, a magnetic force, or any other kind of force.

Now, a computational template cannot simply be picked out from the theoretical template. This is the kind of feature that drives the solving-problem viewpoint, but not the description of patterns of behavior viewpoint. To the latter viewpoint, there is an entire methodology that mediates between the computational model and the theoretical model that needs to be explored. Concretely, the process of construction of a template involves a number of idealizations, abstractions, constraints, and approximations of the target system for which researchers needs to account. Moreover, at some point the computational template needs to be validated against data. What happens when it fails to fit those data? Well, the answer is that researchers have a series of well-established methods for correcting the computational template in order to ensure accurate results. According to Humphreys, the *construction assumptions* and

---

[9]There are many other contemporary authors that deserves our attention. Most prominently is the work of Claus Beisbart, who takes computer simulations as *arguments* (Beisbart 2012). That is, an inferential structure encompassing a premise and a conclusion. Another interesting case is Rawad Swaf and Cyrille Imbert (2013), who conceptualize computer simulations as 'unfolding scenarios.' Unfortunately, space does not allow me to discuss these authors in more extent.

[10]For details, see Humphreys (2004, 102–103).

correction set—components two and three in the sextuple—fulfill precisely these roles. Without them, the computational template might not even be computable.

Now, in order to have an accurate representation of the target system, the variables, functions, and the like in the computational template need to be given an *interpretation*. For example, in the first derivation of a diffusion equation, the interpretation of the function representing the temperature gradient in a perfectly insulated cylindrical conductor is central to the decision on whether the diffusion equation correctly represents the flow of heat in a given metal bar (Humphreys 2004, 80). The researcher's interpretation of the computational template constitute part of the justification for adopting certain equations, values, and functions. The computational templates, says Humphreys, are "not mere conjectures but objects for which a separate justification for each idealization, approximation, and physical principle is often available, and those justifications transfer to the use of the template" (Humphreys 2004, 81).

Finally, the output representation, that is, the visualization of computational model, comes in different flavors. It can be a data array, functions, matrix, and more important in terms of understanding, dynamic representations such as videos or interactive visualizations. As we will discuss in detail in Sect. 5.2.1, visualizations play a fundamental role in our epistemic gain using computer simulations, and thus in their general success as novel methods in scientific and engineering research.

Eric Winsberg is the second philosopher on our list. According to him, there are two core characteristics that meaningfully distinguish computer simulations from other forms of calculation. First, much effort goes into setting up the model that serve as the basis for computer simulations, as well as to deciding which simulation results are reliable and which are not. Second, computer simulations make use of a variety of techniques and methods that facilitate drawing inference from results (Winsberg 2010). As discussed earlier in this section, these two characteristics are typical from the description of patterns of behavior viewpoint.

Furthermore, Winsberg correctly points out that the construction of computer simulations are guided, but not determined, by theory. This means that, although computer simulations rely on theoretical background, they typically encompass elements that are not directly related to, nor are part of, theories. A case of this are 'fictionalizations,' that is, *contrary-to-fact* principles that are included in the simulation model with the purpose of increasing the reliability and trustworthiness of its results. As we saw earlier, Humphreys made a similar point with the construction assumptions and correction set. Winsberg then illustrates fictionalizations with two examples, 'artificial viscosity' and 'vorticity confinement.' In simulations of fluid dynamics, these techniques are successfully used despite not offering realistic accounts of the nature of fluids. Why are they used then? There are several reasons, including of course that they are largely part of the practice of model-building techniques on fluid dynamics. Other reasons include the fact that these fictionalizations facilitate the calculation of crucial effects that would otherwise be lost, and that without these fictionalizations, the results of simulations on fluid dynamics could neither be accurate nor justified.

The previous discussions show that it is simply not possible to fit the concept of computer simulations into one conceptual corset. Thus, our initial question: 'what are computer simulations?' cannot be uniquely answered. It seems that, ultimately, it will depend on the commitments of the practitioners. Whereas the problem-solving viewpoint is more interested in finding solution to complex models, the description of patterns of behavior viewpoint is concerned with accurately representing a target system. Both offer good conceptualizations of computer simulations, and both have several problems to face. Let us next discuss three different kinds of computer simulations found in the scientific and engineering practice.

1.2   Kinds of Computer Simulations

Before addressing the different classes of computer simulations, let us briefly discuss a short classification of target systems typically associated with computer simulations. This classification, besides being non-exhaustive—or precisely because of this—holds no expectations of being unique. Other ways of characterizing target systems—along with the models that represent such target systems—can lead to a new and improved taxonomy.

Having mentioned all the usual warnings, let us begin with the most familiar of all target systems, that is, *empirical target systems*. These are empirical phenomena—or real-world phenomena—in all forms and flavors. Examples include microwave background radiation and Brownian movement in astronomy and physics, social segregation in sociology, competition among vendors in economy, and scramjets in engineering, among many others examples.

Understandably, empirical target systems are the most pervasive target system in computer simulation. This is chiefly because researchers are seriously engaged in understanding the empirical world, and computer simulations provide a new and successful method for achieving such aims. Now, in order to represent empirical target systems, computer simulations implement models that theoretically underpin real-world phenomena with the help of laws, principles, and theories accepted by the scientific community. The Newtonian model of planetary movement, for instance, describes the behavior of any two bodies interacting with each other by a handful of laws and principles. Unfortunately, not every empirical target system can be so simply and accurately represented.

More commonly, computer simulations represent real-world phenomena by including a plethora of elements from different—and sometimes incompatible—sources. Take for instance *scramjets*, combustion ramjets in which combustion takes place in supersonic airflow. The use of Navier-Stoke equations are typically at the base of simulations of fluid dynamics. However, the intake of a scramjet compresses the incoming air via a series of shock waves generated by the specific shape of the intake along with the high flight velocity, as opposed to other air-breathing vehicles that compress the incoming air by compressors—or other moving parts. Simulations of scramjets, then, cannot be fully characterized by Navier-Stoke equations. Instead, the laminar and turbulent boundary layers, along with the interaction with shock waves, yields a three-dimensional unsteady complex flow pattern. A reliable simulation of what is happening within the intake is, then, accomplished by means of high fidelity direct numerical simulations and large eddy simulations. It is the model design, programmed, and built by the engineers, and not just the Navier-Stoke equations, that permits a reliable simulation (Barnstorff 2010).[11]

Another important target system is the so-called *hypothetical target system*. These are target systems where no empirical phenomena are described. Rather, they are either *theoretical* or *imaginary*. A *theoretical target system* describes systems or processes within the universe provided by a theory, whether mathematical (e.g., a torus), physical (e.g., air resistance equal to zero), or biological (e.g., infinite populations). Take as example the famous problem of the Seven Bridges of Königsberg,[12] or the Traveling Salesman Problem.[13] Thus understood, the target system is not empirical, but instead it has the properties of a mathematical or a logical system. A computer simulation implementing these models is mostly theoretical in essence, and it is typically designed for exploring the underlying properties of the model.

Imaginary target systems, on the other hand, stand for non-existing, imagined scenarios. For instance, an epidemic outbreak of influenza in Europe counts as such a target system. This is because such a scenario is prone to never exist, although it does not mean that it will never happen. A simulation of such a scenario provides researchers the necessary understanding of the dynamics of an epidemic outbreak for planning prevention measures and containment protocols, as well as for training personnel. Imaginary systems can be, in turn, divisible into two further kinds, namely, *contingent* and *impossible* (Weisberg 2013). The former stands for a scenario that, as a matter of contingent fact, does not exist. The latter stands for a scenario that is nomologically impossible. The simulation of an epidemic outbreak is a case in point of the former, while running a simulation that violates the known laws of nature is an example of the latter.

The first thing to note about this classification is that computer simulations could represent one target system but render results of another target system. This is a common 'jumping' mechanism that could be harmless, or that could cast a shadow of doubt on the results. A simple example will show how this is possible. Consider simulating the planetary movement implementing a Newtonian model; now instantiate $G = 2\,\mathrm{m}^3\mathrm{kg}^{-1}s^{-2}$ as an initial condition. The example shows a simulation that initially implements an *empirical target system*, but renders results of a *nomologically impossible imaginary target system*. To the best of our current knowledge of

---

[11]I should also mention that there are several other contrivances also involved in the design and programming of computer simulations. In this respect, Sect. 4.2 presents and discusses some of them, such as calibration procedures, and verification and validation methods.

[12]The problem can be best described as finding a way to cross each of the seven bridges of the city of Königsberg only once. The problem, solved by Euler in 1735, laid the foundations of graph theory.

[13]The Travelling Salesman Problem describes a salesman who must travel between N number of cities and keep the travel costs as low as possible. The problem consists in finding the best optimization of the the salesman's route.

the universe, there is no such gravitational constant. As a consequence, the results of a simulation which in principle should have been sanctioned empirically (e.g., by validating against empirical data) can only be confirmed theoretically.[14]

In a way, these issues are part of the general charm and malleability of using computer simulations, but they need to be taken seriously by philosophers and sociologists science. Having said this, and looking closely at the practice of computer simulation, one could sees how researchers have a few 'tricks' that help to cope with situations like 'jumping'. For instance, one solution to the simulation of the satellite under tidal stress would be to set G as a global constant of value $6.67384 \times 10^{-11} \mathrm{m}^3 \mathrm{kg}^{-1} \mathrm{s}^{-2}$. Unfortunately, this is only a palliative solution since the value of the variable *mass of satellite* could be set to any unrealistically large value. Again, researchers could establish lower and upper limits on the size of the satellite and planet mass, but this solution only begs the question if there is not another way to 'fool the simulation'.

Either by modeling or by instantiation, computer simulations can create several scenarios out of the mind of researchers. How do researchers turn this seemingly disastrous situation into something advantageous? The answer is, I believe, in the way scientists and engineers sanction computer simulations as reliable processes. That is, by providing reasons for believing that computer simulations are a reliable process which renders, most of the time, correct results. I will explore these issues in more detail in Chap. 4.

Along with a classification of target systems, I now suggest a classification of computer simulations. In the same spirit, this classification is not meant to be exhaustive, conclusive, nor unique. Here, I divide simulations into three classes, based on the standard treatment that computer simulations have received from the specialized literature (Winsberg 2015). These are *cellular automaton, agent-based simulations,* and *equation-based simulations.*

1.2.1   Cellular Automata

Cellular automata are the first of our examples of computer simulations. They were devised in the 1940s by Stanislaw Ulam and John von Neumann while Ulam was studying the growth of crystals using a simple lattice network as a model, and von Neumann was working on the problem of self-replicating systems. The story goes that Ulam suggested to von Neumann to use the same kind of lattice network as his, creating in this way a two-dimensional, self-replicator algorithm.

Cellular automata are simple forms of computer simulations. Such simplicity stems from both their programming and underlying conceptualization. A standard cellular automaton is an abstract mathematical system where space and time are considered to be discrete; it consists of a regular grid of cells, each of which can be in any state at a given time. Typically all the cells are governed by the same rule,

---

[14]Running a second computer simulation that could confirm these results is becoming standard practice (Ajelli et al. 2010).

which describes how the state of a cell at a given time is determined by the states of itself and its neighbors at the preceding moment. Stephen Wolfram defines cellular automata in the following way:

> [...] mathematical models for complex natural systems containing large numbers of simple identical components with local interactions. They consist of a lattice of sites, each with a finite set of possible values. The value of the sites evolve synchronously in discrete time steps according to identical rules. The value of a particular site is determined by the previous values of a neighborhood of sites around it. (Wolfram 1984b, 1)

Although a rather general characterization of this class of computer simulation, the above definition already provides the first ideas as to their domain of applicability. Cellular automata have been successfully used for modeling many areas in social dynamics (e.g., behavioral dynamics for cooperative activities), biology (e.g., patterns of some seashells), and chemical types (e.g., the Belousov-Zhabotinsky reaction).

One of the simplest and most canonical examples of cellular automata is Conway's *Game of Life*. The simulation is remarkable because it provides a case of emergence of patterns and self-organization dynamics of some systems. In this simulation, a cell can only survive if there are either two or three other living cells in its immediate neighborhood. Without these companions, the rule indicates that the cell dies either from overcrowding, if it has too many living neighbors, or from loneliness, if it has too few. A dead cell can make its way back to life provided that there are exactly three living neighbors. In truth, there is little interaction—as one would expect from a game—besides creating an initial configuration and observing how it evolves. Nevertheless, from a theoretical point of view, the Game of Life can compute any computable algorithm, making it a remarkable example of a universal Turing machine. Back in 1970, Conway's Game of Life open up a new field of mathematical research: the field of cellular automata (Gardner 1970).

Elementary cellular automata furnishes some fascinating cases in contemporary science. The idea of these automata is that they are based on an infinite one-dimensional array of cells with only two states. At discrete time intervals, every cell changes state based on its current state and the state of its two neighbors. Rule 30 is a case in point which produces complex, seemingly random patterns from simple, well-defined rules (see Fig. 1.4). For instance, a pattern resembling Rule 30 appears on the shell of cone snail species *Conus textile* (see Fig. 1.5). Other examples are based on it's mathematical properties, such as using Rule 30 as a random number generator for programming languages, and as a possible stream cipher for use in cryptography. The rule set which governs the next state of Rule 30 is shown in Fig. 1.4.

Cellular automata entrench a set of unique methodological and epistemological virtues. To name a few, they accommodate better to error because they render exact results of the model they implement. Since approximations with the target system are almost nonexistent, any disagreement between the model and the empirical data can be ascribed directly to the model that realized the set of rules. Another epistemological virtue is pointed out by Evelyn Fox-Keller, who explains that cellular automata lack theoretical underpinning in the familiar sense of the term. That is, "what is to

current pattern	111	110	101	100	011	010	001	000
new state for center cell	0	0	0	1	1	1	1	0

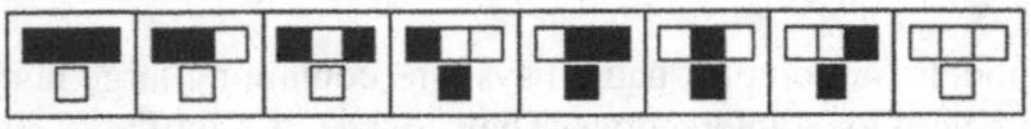

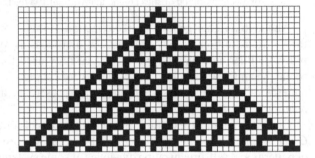

Fig. 1.4 Rule 30. http://mathworld.wolfram.com/CellularAutomaton.html

Fig. 1.5 Pattern created by
Rule 30. http://mathworld.
wolfram.com/
CellularAutomaton.html

be simulated is neither a well-established set of differential equations [...] nor the fundamental physical constituents (or particles) of the system [...] but rather the phenomenon itself" (Fox Keller 2003, 208). Approximations, idealizations, abstractions and the like are concepts that worry the practitioner of cellular automata very little.

Now, not everything is great for cellular automata. They have been criticized on several grounds. One of these criticisms touches upon the metaphysical assumptions behind this class of simulation. It is not clear, for instance, that the natural world is actually a discrete place, as assumed by the cellular automata. Many of today's scientific and engineering work is based on the description of a continuous world. On less speculative grounds, it is a fact that cellular automata have little presence in scientific and engineering fields. The reason for this, I believe, is partly cultural. The physical sciences are still the accepted viewpoint for describing the natural world, and they are written in the language of Partial Differential Equations and Ordinary Differential Equations (PDE and ODE respectively).

Naturally, advocates of cellular automata focus their efforts to show their relevance. In all fairness, many cellular automata are more adaptable and structurally similar to empirical phenomena than PDEs and ODEs (Wolfram 1984a, vii). It has been pointed out by Annick Lesne, a renown theoretical physicist, that discrete and continuous behavior coexist in many natural phenomena, depending on the scale of observation. To her mind, this is an indicator not only of the metaphysical basis of many natural phenomena, but also of the suitability of cellular automata for scientific and engineering research (Lesne 2007). In a similar vein, Gérard Vichniac believes that cellular automata seek not only numerical agreement with a physical system, but also they attempt to match the simulated system's own structure, its topology, its symmetries and its 'deep' properties (Vichniac 1984, 113). Tommaso Toffoli has a similar stand as these authors, to the point that he entitled a paper: "Cellular automata as an alternative to (rather than an approximation of) differential equations in model-ing physics" (Toffoli 1984), highlighting cellular automata as the natural replacement of differential equations in physics.

Despite these and many other authors' efforts to show that the world might be more adequately described by cellular automata, the majority of scientific and engi-neering disciplines have not made a complete shift yet. Most of the work done in these disciplines is predominantly based on agent-based and equation-based simulations. As mentioned before, in the natural sciences and engineering, most physical and chemical theories used in astrophysics, geology, climate change, and the like imple-ment PDE and ODE, two systems of equations that are at the basis of equation-based simulations. Social and economic systems, on the other hand, are better described and understood by means of agent-based simulations.

1.2.2 Agent-Based Simulations

While there is no general agreement on the definition of the nature of an 'agent', the term typically refers to self-contained programs that control their own actions based on the perceptions of the operating environment. In other words, agent-based simulations 'intelligently' interact with their peers as well as their environment.

The relevant characteristic of these simulations is that they show how the total behavior of a system emerges from the collective interaction of their parts. To decon-struct these simulations into their constituent elements would remove the added value that has been provided in the first place by the computation of the agents. It is a fun-damental characteristic of these simulations, then, that the interplay of the various agents and the environment brings about a unique behavior of the entire system.

Good examples of agent-based simulations proceed from the social and behavioral sciences, where they are heavily present. Perhaps the most well-known example of an agent-based simulation is Schelling's Model of Social Segregation.[15] A very simple description of Schelling's model consists in two groups of agents living in a 2-D,[16] $n \times m$ matrix 'checkerboard' where agents are placed randomly. Each individual agent has a *3 x 3* neighborhood, which is evaluated by a utility function that indicates

---

[15]Although nowadays Schelling's model is carried out by computers, Schelling himself warned against its use for understanding the model. Instead, he used coins or other elements to show how segregation occurred. In this respect, Schelling says: "I cannot too strongly urge you to get the nickels and pennies and do it yourself. I can show you an outcome or two. A computer can do it for you a hundred times, testing variations in neighborhood demands, overall ratios, sizes of neighborhoods, and so forth. But there is nothing like tracing it through for yourself and seeing the process work itself out. It takes about five minutes no more time than it takes me to describe the result you would get" (Schelling 1971, 85). Schelling's warning against the use of computers is an amusing anecdote that illustrates how scientists could sometimes fail in predicting the role of computers in their own respective fields.

[16]Schelling also introduced a 1-D version, with a population of 70 agents, with the four nearest neighbors on either side, the preference consists of not being minority, and the migration rule is that whoever is discontent moves to the nearest point that meets her demands (Schelling 1971, 149).

Fig. 1.6 Initial random
distribution of agents in a
checkerboard 13 rows × 16
columns, with a total of 208
squares (Schelling 1971,
155)

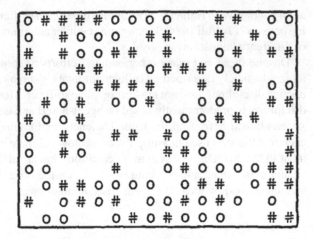

the migration criteria. That is, the set of rules that indicates how to relocate—if possible—in case of discontent by an agent (see Fig. 1.6)

Schelling's segregation model is a canonical example of agent-based simulations.[17] But more complex agent-based simulations can be found in the literature. It is now standard that researchers model different attributes, preferences, and overall behavior into the agents. Nigel Gilbert and Klaus Troitzsch list the set of attributes that are typically modeled by agents (Gilbert et al. 2005):

1. *Knowledge and belief*: Since agents base their actions on their interaction with their environment as well as other agents, it is crucial to be able to model their system of beliefs. The traditional distinction between *knowledge*, as true justified belief, and mere *belief* can then be modeled. Those whose information might be faulty or plain false must be modeled to act in their environment in a rather different way than those agents whose information is correct, as a result of knowing.

2. *Inference*: Knowledge and belief are possible because agents are able to infer information from their own set of beliefs. Such inferences are modeled in different ways, sometimes over very intuitive assumptions. For instance, agent *A* could infer that a source of 'food' is near agent *B* knowing—or simply believing—that agent *B* has 'eaten some food.'

3. *Goals*: Since agents are, for the most part, programed as autonomous entities, they are typically driven by some sort of goals. Survival is a good example of this, as it might require the satisfaction of subsidiary goals, such as acquiring sources of energy, food, and water, as well as avoiding any dangers. Modeling these subsidiary goals is not easy, since researchers must decide on how to weigh the importance and relevance among several goals. Different design decisions lead to different goal-guided agents, and therefore to a different overall behavior

---

[17]In truth, depending on how the Schelling model is design and programmed, it could also qualify as an Cellular Automata. Thank you Andrés Ilčić for pointing this out to me.

of the system. Let us note that the goals of an agent are a different attribute from knowing and inferring. Whereas the former are about guiding the general behavior of the agents in their environment, the latter depends on valid or true information for ruling their behavior.

4. *Planning*: In order to satisfy its goals, an agent needs to have some way of determining what decisions are best. Typically, a set of condition-action rules are programed as constituent of the agent. For instance, a utility function is programed for the satisfaction of an energy source and for what counts as a 'danger.' Planning involves inferring what actions lead to a desired goal, what state is required before that action takes place, and what actions are needed to arrive at the desired state. In this respect, planning is very sophisticated as the agent needs to weigh several options, including having a 'pay-off' rule over their decisions, determining where it needs to be at some point in the future, and the like. It has been objected that agent planning is not realistic of human planning because most human action is driven by routine decisions, an inherent tendency to discern, and even instinctive judgments that cannot be modeled by a calculated plan.

5. *Language*: Passing information across agents is central for any agent-based simulation. An exciting example is Alexis Drogoul and Jacques Ferber's multi-agent model of ants' colonies (Drogoul et al. 1992). According to the authors, agents can communicate by propagating a 'stimuli' to the environment. This stimuli might be received and transmitted in different ways. When ants receive this stimuli, they activate a friendly behavior. However, when a predator receives the stimuli, it triggers an aggressive behavior. In this particular agent-based simulation, such a communication mechanism is very simple and by no means aims at conveying any meaning. Other examples of modeling language in agent-based simulations are the negotiation of contracts (Smith and Davis 1981), communication of decisions, and even one agent threatening another with 'death' (Gilbert et al. 2005).

6. *Social models*: Some agent-based simulations, like Schelling's segregation model discussed earlier, aim at modeling the interrelationship between agents in a larger environment. On this account, agents are able to create their own topology given the set of rules, their interaction with other agents, and the initial setup, among other parameters. For instance, in Schelling's segregation model, the agents create different topologies of segregation given the side of the grid, the utility function, and the initial position of the agents.

1.2.3   Equation-Based Simulations

Perhaps the most commonly used class of computer simulations in science and engineering are the so-called equation-based simulations. At their basis, they are the implementation of a mathematical model on the physical computer which describes a target system. Because most of our understanding of the world comes through the use of mathematical descriptions, these simulations are by far the most popular in scientific and engineering fields. Naturally, examples abound. Fluid dynamics, solid mechanics, structural dynamic, shock wave physics, and molecular chemistry, are just a handful of cases. William Oberkampf, Timothy G. Trucano, and Charles Hirsch have elaborated on an extensive list that they deem to call 'computational engineering' and 'computational physics' (Oberkampf et al. 2003). Let us note that their labeling emphasizes simulations *only* in physics and engineering fields. Although it is correct to say that the overwhelming majority of equation-based simulations can be found in these domains, it does not do justice to the myriad of simulations also found in other scientific fields. To extend the examples, the Solow-Swan model of economic growth is a case in economy, and the Lotka-Volterra model of predator-prey works for sociology as well as for biology.

As mentioned, this class of simulations implement mathematical models on the computer. But, is that simply so? From Sect. 1.1 we learned that this way of thinking corresponds to the problem-solving viewpoint of computer simulation. According to its advocates, there is little that mediates between the mathematical model and its implementation on the computer as a computer simulation. The opposite view, the decription of pattern of behavior viewpoint, takes it that there is in fact a methodology that facilitates the implementation of—a multiplicity of—mathematical models into the computer. To have a better grasp of a typical equation-based simulation—and to determine which viewpoint is closer to the actual practices—, let us briefly analyze an example of a recent simulation on the age of the Sahara desert.

Zhongshi Zhang et al. believed that the Sahara desert emerged during the Tortonian stage—approximately 7–11 million years ago—of the Late Miocene epoch after a period of aridity in the north African region (Zhang et al. 2014). To prove their hypothesis, Zhang's team decided to simulate the climate change in these regions on geological timescales and over the past 30 million years. The Sahara's age, according to the simulation is, in effect, of about 7 to 11 million years old. With this result in hand, Zhang's et al. were able to oppose most of the standard estimations of the age of the Sahara, which take it to be about 2–3 million years old at the onset of the Quaternary ice ages. How did they actually simulate such a complex empirical system?

First, the authors do not implement one grand model of climate change on the computer and calculate it until obtaining the results. This is the problem-solving viewpoint that understand computer simulations solely as means for computation. Computer simulations are much like a laboratory workbench, where scientists subtly combine pieces of theory, with bits of data, and a lot of expertise knowledge and instinct. The process is in fact complex, messy, and in many cases non-standardized. Zhang's team made use of a family of models, each performing different tasks and representing a different aspect of the target system. They made use of low- and high-resolution versions of the Norwegian Earth System Model (NorESM-L) for accounting for the series of geologic epochs, and the Community Atmosphere Model version 4 (CAM4) as the atmosphere component of NorESM-L. In fact, the NorESM-L model is in itself a hierarchy of small models—or mere components of a larger model—representing the land, sea ice, the ocean, etc.

There is, in truth, no grand model that could tell us about the age of the Sahara. Instead, a patchwork of models—some known and well-established, some speculative—laws, principles, data, and bits of theory is what conform Zhang team's simulation. This should come as no surprise, as it is generally assumed that there is no general theory that either underpins or guides computer simulations. Moreover, simulations typically include non-linguistic information, such as design decisions, possible model bias, identified uncertainties, and "not included in this model" disclaimers. Bentsen et al., when describing CAM4, furnish a good example of such a disclaimer: "[a]erosol indirect effects on mixed-phase and ice clouds (e.g. (Hoose et al. 2010) are not included in the current version of CAM4-Oslo" (Bentsen et al. 2013), 689).

Despite their lack of full theoretical underpinning, these simulations are still highly reliable as they represent a specific target system and are typically validated by standard verification and validation methods (see Sect. 4.2.2). In this respect, Zhang et al. are constantly reminding us that the model performs well in simulating the pre-industrial climate, that CAM4 simulate the patterns of modern African rainfall reasonably well, and other similar confirmatory stances of the models. Such reminders, of course, cannot stop objections against the results of the simulation. In particular, critics of Zhang's work point out the lack of evidence for validating their results. Stefan Kröpelin is a chief detractor of using computer simulations for these kinds of target system. He admits that, although the model is interesting, it is mainly "numerical speculation based on almost non-existent geological evidence (...) Nothing you can find in the Sahara is older than 500,000 years, and in terms of Saharan climate even our knowledge of the past 10,000 is full of gaps" (Kroepelin 2006). The response by Zhang et al. is that the evidence for the early onset of the Saharan aridity is highly contentious. Mathieu Schuster also disagrees with Kröpelin's interpretation of data. According to him, "although it is true that too little is known about the ancient geology of the region [...] the 2006 Chad study [...] as well as the ones that reported increases in dust and pollen from sediment, contained 'strong pieces of evidence to support our new findings'" (Schuster 2006). In fact, the simulation by Zhang and his team come to support some claims already in the literature. Anil Gupta and his team claim an increased upwelling in the Indian Ocean about 7–8 million years ago (Gupta et al. 2004); and Gilles Ramstein and his team used modeling experiments to show Eurasian summer temperatures increase in response to the Tethys shrinkage, which would also enhance the monsoon circulation (Ramstein et al. 1997).

To state that computer simulations are unreliable, or that their results do not correlate to the way the world is requires more than just a claim that there is no 'evidence' that supports the unreliability of the simulation.[18] Other indicators of reliability play a central role as well. For instance, the ability of the simulation to explain and predict direct or related phenomena. According to Zhang et al., their simulation shows that the African summer monsoon was drastically weakened by the

---

[18]This is especially true for the traditional sense of 'evidence' (i.e., empirically based) that Kröpelin refers to. Other forms of evidence also include results of well-established simulations, verification and validation, convergence of solutions, etc.

Tethys Sea shrinking during the Tortonian stage, allowing the alteration of the mean climate of the region. Such climatic change, the researchers speculate, "probably caused the shifts in Asian and African flora and fauna observed during the same period, with possible links to the emergence of early hominis in north Africa" (Zhang et al. 2014, 401). Interestingly enough, researchers could reach such a conclusion only by means of running computer simulations.

Allow me to finish this section with a short description of the general computational methods for solving the equation-based simulations. Depending on the problem and the availability of resources, one or more of the following methods apply: *analytical methods*, *numerical analysis* and *stochastic techniques*.

- *Exact solutions:* this is the simplest method of all. It consists of carrying out the operations specified in the simulation in a similar fashion as a mathematician would do using pen and paper. That is, if the simulation consist in adding $2 + 2$, then the result must be 4—as opposed to an approximate solution. Computers have the same capacity to find the exact solutions to certain operations as any other computational mechanism, including our brain. The efficacy of this method depends, however, on whether the size of the 'word' in a computer is large enough for carrying out the mathematical operation.[19] If the operation exceeds its size, then round-off and truncation mechanisms intervene for the operation to be possible, though with a loss in accuracy.
- *Computer-implemented numerical methods*: this method refers to computer-implemented methods for calculating the simulation model by approximation. Although mathematical studies on numerical analysis predate the use of computers, they gain importance with the introduction of computers for scientific and engineering purposes. These methods are used for solving PDE and ODE equations, and include linear interpolation, the Runge-Kutta method, the Adams-Moulton method, Lagrange interpolation polynomial, Gaussian elimination, and Euler's method, among others. Let it be noted that each method is used for solving a specific kind of PDE and ODE, depending on how many derivatives involve the unknown function of n variables.
- *Stochastic techniques:* for higher order dimensions, both exact solutions and computer-implemented numerical methods become prohibitively expensive in terms of computational time and resources. Stochastic techniques rely on methods that use pseudo-random numbers; that is, numbers generated by a numerical engine.[20] The most famous stochastic method is the Monte Carlo method, which is particularly useful for simulating systems with many coupled degrees of freedom such as fluids, disordered materials, strongly coupled solids, and cellular structures, to mention a few.

---

[19] A 'word' represents the minimum unit of data used by a particular computer architecture. It is a fixed sized group of bits that are handled as a unit by the processor.

[20] The prefix 'pseudo' reflects the fact that these methods are based on an algorithm that produces numbers on a recursive basis, eventually repeating the series of numbers produced. Pure randomness in computers can never be achieved.

1.3 Concluding Remarks

This chapter had the sole purpose of addressing the question 'what is a computer simulation?' This is of course an important question, since it sets the grounds for much of what is discussed about computer simulations later in this book. For this reason, the first part of the chapter deals with some historical remarks about the many attempts to define computer simulations, whether offered by engineers, scientists or philosophers. In this context, I distinguished two kinds of definitions. Those that emphasize the computing power of computer simulations—called the *problem-solving technique* viewpoint—and those that take computer simulations to have, as a chief feature, the capacity to represent a given target system—called *description of patterns of behavior* viewpoint. Although there are a handful of definitions where both viewpoints are combined, and arguably one that does not fit with our distinction, in general, researchers across fields agree on conceptualizing computer simulations as one or the other viewpoint.

The second part of this chapter dealt with three different kinds of computer simulations, as standardly found in the literature. These are, *cellular automata, agent-based simulations*, and *equation-based simulations*. As warned, this is neither an exhaustive taxonomy nor offers a unique classification. It could be relatively simple to show how an agent-based simulation could be interpreted as cellular automata (e.g., when focused on their nature as agents/cells), or as an equation-based simulation (e.g., if the inner structure of an agent is equations). The key is to see which characteristic of the computer simulation is highlighted. Here, I offered some criteria for a sound characterization of each type. A final warning was issued, however, regarding the methodology and epistemology tailored to each kind. It is not difficult to show that each kind of computer simulation entails specific and distinct methodological and epistemological concerns, and therefore they require a different treatment in their own way. In the remainder of this book, I focus my attention solely on the so-called equation-based simulations.

Chapter 2
Units of Analysis I: Models and Computer Simulations

Theories, models, experimental set-ups, prototypes: these are some of the typical units of analysis found in standard scientific and engineering work. Science and engineering are of course populated with other, equally decisive units of analysis that facilitate our description and knowledge of the world. These include hypotheses, conjectures, postulates, host of theoretical machinery. Computer simulations are the new acquisition in the scientific and engineering arena that count as novel units of analysis.[1] What are the constituents of computer simulations that comprise such a new unit? What makes them different from other units of analysis? Answers to these questions are put forward here.

In the previous chapter, I turned our interests to computer simulations that implement equations as regularly found and used in the sciences and engineering. This chapter aims at making these general remarks more specific. To this end, the first section clarifies the notion of scientific and engineering model, as it is the basis of equation-based simulations. I also mentioned that their implementation is not directly on the physical computer—recall that this was a fundamental assumption of the description of patterns of behavior viewpoint—but rather mediated by a proper methodology for computer simulations. The second part of this chapter presents in some detail how models are implemented as 'simulation models.' To this end, I present and discuss three main constituent parts of computer simulations, namely, *specifications*, *algorithms*, and *computer processes*. I will then present important social, technical, and philosophical problems emerging from this characterization. By doing this, I hope, we will be entrenching the status of computer simulations as units of analysis in their own right.

---

[1] Big Data—which I discuss in Sect. 6.4—and Machine Learning should also be included as novel units of analysis in science and engineering.

© Springer Nature Switzerland AG 2018
J. M. Durán, *Computer Simulations in Science and Engineering*,
The Frontiers Collection, https://doi.org/10.1007/978-3-319-90882-3_2

2.1　Scientific and Engineering Models

Todays scientific and engineering research heavily depends on models. But what is a 'scientific model'? and what is an 'engineering model'? are they different? and how can they be studied? On the face of it, these questions can be answered in many different ways. Tibor Müller and Harmund Müller have provided nineteen enlightening examples of the different ways in which the notion of model is found in the literature (Müller and Müller 2003, 1–31). Some notions emphasize the uses and purposes of models. An example is Canadarm 2—formally the Space Station Remote Manipulator System (SSRMS)—which was first conceived as a model for assistance and maintenance on board the International Space Station and then became the robotic arm that is known today. Some other notions are more interested in understanding the epistemological input of a model. For instance, a Newtonian model provides insight into the planetary movement, whereas a Ptolemaic model fails to do so. Furthermore, some definitions highlight the propaedeutic value of models above everything else. In this latter case, the Ptolemaic model is as valuable as the Newtonian model since they both exemplify different scientific standards.

A standard way to devise models depends on their material properties—or lack thereof. Call *material models* those models which are physical in a straightforward sense, such as models made of wood, steel, or any other kinds of materials (e.g., Canadarm); and *conceptual* or *abstract models* those models which are the product of abstraction, idealizations, or fictionalizations of a target system, such as theoretical models, phenomenological models, and data models, among other kinds (e.g., the Newtonian and Ptolemaic model of the planetary system).[2]

At first sight, material models seem closer to the notion of laboratory experimentation than abstract models, which are in turn closer to computer simulations. As we shall discuss in Chap. 3, some philosophers have used this distinction to ground their claims about the epistemological power of computer simulations. Specifically, it has been argued that the abstractness of computer simulations reduces their capacity to infer back to the world, and thus they are epistemically inferior to laboratory experimentation. As it turns out, however, there is just as much abstraction involved in laboratory experimentation as there is in computer simulations—although, naturally, not to the same degree. The converse, on the other hand, is not true. Computer simulations do not have any use for material models, as all they can implement are conceptual models. Let us now briefly discuss these ideas.

Material models are sometimes intended to be a 'piece' of the world and, as such, a more or less accurate replica of an empirical target system. Take as example the use of a beam of light for understanding the nature of light. In such a case, the material model and the target system share obvious properties and relations: the model and the target system are made of the same materials. In this particular case, the distinction between a model of a beam of light and a beam of light in itself is only programmatic: there are no real differences except for the fact that it is easier for the researcher to manipulate the model than a real beam of light. This fact is especially true when there

---

[2]For an excellent treatment on models and computer simulations, see the work of Morrison (2015).

are differences of scale between the model and the target system. In some cases, it is much easier to turn on a flashlight than trying to capture some sunlight coming through the window.

Of course, material models do not always need to be made of the *same* materials as the target system. Consider for instance the use of a ripple tank for understanding the nature of light. The ripple tank is a material model in a straightforward sense, as it is made of metal, water, sensors, etc. However, its basic material setup differs significantly from its target system, which is light. What could lead scientists to believe that the ripple tank could help them to understand the nature of light? The answer is that waves are easily reproduced and manipulated within a ripple tank, and therefore, it is very useful for understanding the wave nature of light.

Here we have an example of a material model (i.e., the ripple tank) that involves certain levels of abstraction and idealization with respect to its target system. The argument is simple. Since the medium in which waves travel in the ripple tank (i.e., water) is different from the target system (i.e., light), there must be a high-level abstract representation that links those two together. This comes along with Maxwell's equations, D'Alambert's wave equation, and Hook's law. It is because the two systems (i.e., water waves and light waves) obey the same laws and can be represented by the same set of equations that researchers can use a material model of one kind to understand a materially different kind.

Now, regardless of the many ways in which material models can be interpreted, they can never be implemented on a computer. As expected, the reason is purely based on their materiality: only conceptual models are suitable as computer simulations. Conceptual models are an abstract—and sometimes formal—representation of a target system and, as such, they are on the opposite ontological side of material models. This means that they are intangible, they are not bound to deterioration—although they are forgettable—and in a sense they exist outside time and space just like mathematics and the imagination.

There is general agreement to posit these kinds of models as interpreted structures that facilitate thinking about the world.[3] Such a structure incorporates a broad range of elements, including equations, theoretical terms, mathematical concepts and techniques, policy views, and metaphors, among others.[4] In addition, philosophers also agree that diverse levels of abstractions, idealizations, and approximations are attached to these models as part of their inherent structures.[5] An example will illumi-

---

[3]For two excellent books on scientific models, see Morgan and Morrison (1999) and Gelfert (2016).

[4]Scientific models and theories sometimes include terms with no specific interpretation, like 'black hole', 'mechanism', and the like. These are known as *metaphors*, and are typically used for filling the gap introduced by these terms. Their use is to inspire some kind of creative response in the users of the model that cannot be rivaled by literal language. Moreover, there is a special relation between metaphorical language and models, one that includes subtleties about models *as* metaphors (Bailer-Jones 2009, 114). Here, I have no interest in exploring this side of modeling.

[5]The philosophical treatment of *abstraction*, *idealizations*, and *approximations* is fairly similar across the literature. Abstraction aims at ignoring concrete features that the target system possesses in order to focus on their formal set-up. Idealizations, on the other hand, come in two flavors: Aristotelian idealizations, which consist in 'stripping away' properties that we believe not to be

nate some of the terms used so far. Consider a mathematical model of the planetary movement. Physicists begin by bringing together some theories (e.g., a Newtonian model) that already abstract and idealize the planetary movement, bits of empirical evidence (e.g., data gathered from observation) which are pre-selected and post-processed, and as such subject to a series of methodological, epistemological and ethical decisions, and finally, a story that conceptualizes and structures the model as a whole.

Because of its simplicity and elegance, this way of understanding models is favored by a wide number of researchers across disciplines. Let us note the strong presence of a representational connection with the target system. Models abstract, idealize, and approximate because their function is to represent a target system as accurately as possible, leaving out unnecessary details. Philosophers have systematically called attention to the overwhelming need for models to represent. The reasons, it is believed, stem from our notion of progress of science. That is, science would be able to better advance understanding of the world with models that carry out some epistemological function (e.g., explanation, prediction, confirmation). Such epistemological functions, in turn, are understood as holding representational links to the world.

Non-representational models, that is, models that fulfill roles other than to represent a target system—such as a propaedeutic, pragmatic, and aesthetic roles—are, however, rapidly acquiring more prominence in the philosophy of models.[6] In here, however, we are mostly interested in models that represent, more or less accurately, the intended target system. The reason is partly because most practice and theory of computer simulations is still carried out with models that represent a target system, and partly because there are few studies on the non-representational side of computer simulations. Having said this, three kinds of models based on their representational capacity emerge, namely, *phenomenological models*, *models of data*, and *theoretical models*.[7]

A standard example of *phenomenological models* is the liquid drop model of the atomic nucleus. This model describes several properties of the atomic nucleus, such as surface tension and charge, without actually postulating any underlying mechanism. Here is the key characteristic of phenomenological models: they mimic observable properties rather than advancing on a structure underpinning the target system. Such a characteristic should not suggest, however, that certain aspects of phenomenological models could not be derivable from theory. Many models incorporate principles and

---

relevant for our purposes; and Galilean idealizations, which involve deliberate distortions. As for approximations, they are an inexact representation of a feature of the target system, for reasons such as practical limitations in approaching a system, or tractability.

[6]See, for instance, the work of Ashley Kennedy on the explanatory role of non-representational models (Kennedy 2012) and of Tarja Knuuttila on the material dimension of models that makes them objects of knowledge and enables them to mediate between different people and various practices (Knuuttila 2005).

[7]It goes without saying that these three models do not exhaust either all the models found in science and engineering, or the many dimensions philosophers use for analyzing the notion. They are, however, a fairly good characterization of the kind of models suitable for a computer simulation.

laws associated with theory, while remaining phenomenological. The liquid drop model is again the example. While hydrodynamics accounts for surface tension, electrodynamics account for charge. The liquid drop model, nevertheless remains a phenomenological model.

Why would scientists and engineers be interested in such models? One reason is that mimicking the features of a target system is sometimes easier to handle than the theory of such a target system. Fundamental theories such as quantum electrodynamics for condensed matter, or quantum chromodynamics (QCD) for nuclear physics are more accessible when a phenomenological model is used instead of the theory itself.[8]

An interesting fact about phenomenological models is that researchers have sometimes downplayed their place in scientific and engineering disciplines. To many, a fundamental reason for using models is that they postulate underlying mechanisms of phenomena, facilitating in this way making meaningful assertions about the target system. Models that merely describe what we observe, such as phenomenological models, do not carry such a fundamental property. Fritz London, one of the brothers that developed the London-London phenomenological model of superconductivity, insisted that their model must be considered only a *temporary* replacement until a theoretical approximation could be elaborated.

Similar to phenomenological models are *models of data*. They both share the lack of theoretical underpinning, and thus only capture observable and measurable features of a phenomenon.[9] Despite this similarity, there are also differences that make a model of data worth studying on its own right. To begin with, phenomenological models are based on the assessment of the behavior of the target system, whereas models of data are based on the reconstructed data collected from observing and measuring properties of interest about the target system. Another important difference lies in their methodology. Models of data are characterized by a collection of well-organized data, and therefore design and construction differ from phenomenological models. In particular, models of data require more statistical and mathematical machinery than any other model because the collected data need to be filtered for noise, artifacts, and other sources of error.

As a result, the set of problems surrounding models of data are significantly different from other kinds of models. For instance, a standard problem is to decide which data need to be removed and under what criteria. A related problem is to decide which curve function represents all the cleaned data. Is it going to be one curve or several curves? And what data points should be left out of consideration when no curve fits them all? Issues regarding fitting the data into a curve are typically dealt with by statistics inference and regression analysis.

Many examples of models of data come from astronomy, where it is common to find collections of large amounts of data obtained from observing and measuring astronomical events. Such data are classified by specific parameters, such as bright-

---

[8] For reasons why this is the case, see Hartmann (1999, 327).

[9] As we shall see in Sect. 6.2, researchers are making important efforts to find structure behind large amounts of data.

ness, spectrum, celestial position, time, energy, and the like. A usual problem for the astronomer is to know which model can be constructed from a given pile of data. As an example, take the virtual observatory data model, a worldwide project where meta-data are used for classification of observatory data, construction of models of data, and simulation of new data.[10]

Consider now a different kind of model, one that postulates an underlying theoretical mechanism of the target system, and call it *theoretical model*. Traditional exponents of this kind of model are Navier-Stoke models of the dynamics of fluids, Newton's equations of planetary motion, and Schelling's model of segregation. Thus understood, theoretical models embody knowledge that comes from well-established theories and underpin the deep structures of the target system. For several reasons, these kinds of models have been the favorites among researchers. However, as data models and phenomenological models gain terrain with the new technologies, they are epistemically positioned on a par with theoretical models.

Finally, models in engineering sometimes receive a different treatment than models in the sciences. The reason for this is because they are conceived as models for 'doing' rather than for 'representing.' Unfortunately, this dichotomy obscures the fact that all kinds of models are used with some purpose in mind, and in that respect they are as much for 'doing' as they are for 'representing.' Moreover, thinking in this way presupposes a distinction between science and engineering, one which in the field of computer simulations seems to be very much dissolved. Where does sociology end and engineering begin in a simulation implementing the Schelling model of segregation? Sure, there are the sciences, and there are the engineerings, and in many cases those two broad disciplines can be well differentiated.[11] But there are also plenty of intersections. Computer simulations, I believe, lie in one such intersection. Because of this, and for the rest of this book, I treat scientific models and engineering models in a similar way.[12]

---

[10] Among international groups working on observatory models of data, there is the International Virtual Observatory Alliance (Alliance 2018); and the Analysis of the Interstellar Medium of Isolated Galaxies (AMIGA) (interstellar Medium of Isolated GAlaxies 2018).

[11] This is particularly true for material models. For instance, the conceptual model of Canadarm included notions that were distinctively from engineering and distinctively from the sciences. But much of it was a mixture of both. However, it is with the mechanical arm that it becomes unmistakably an engineering artifact. On this point is where much of heavy work in the philosophy of technology is being carried out.

[12] More on modelling in engineering can be found in (Meijers 2009), specially *Part IV: Modelling in Engineering Sciences*

2.2   Computer Simulations

2.2.1   *Constituents of Computer Simulations*

In the previous chapter, I narrowed down the class of computer simulations of interest to equation-based simulations. William Oberkampf, Timothy G. Trucano, and Charles Hirsch call attention to these kind of computer simulations in 'computational engineering and physics,' which includes computational fluid dynamics, computational solid mechanics, structural dynamics, shock wave physics, and computational chemistry, among others.[13] The reason for labeling computer simulations in this way is to emphasize their dual dependence. On the one hand, there is a dependence on the natural sciences, mathematics, and engineering. On the other hand, they rely on computers, computer science, and computer architecture. Equation-based simulations of course also extend to other fields of research as well. In the life sciences, for instance, equation-based simulations are important in synthetic biology, and in medicine there has been an incredible progress with in-silico clinical trials. One could also mention the many simulations performed in the field of economics and sociology.

Philosophers have largely acknowledged the importance of studying the methodology of computer simulations for understanding their place in scientific and engineering research. Eric Winsberg is of the opinion that the credibility of computer simulations does not come solely from the credentials supplied to it by theory, but also and perhaps to a large extent from established credentials in the model-building techniques employed in the construction of the simulation.[14] Similarly, Margaret Morrison takes that representational inaccuracies between simulation models and mathematical models can be resolved at different modeling stages, such as calibration and testing.[15] Furthermore, Johannes Lenhard has cogently argued that the process of simulation modeling takes the form of an explorative cooperation between experimenting and modeling (Lenhard 2007). I am convinced that Winsberg, Morrison and Lenhard are right in their interpretations. Winsberg is right in pointing out that the reliability of computer simulations has diverse sources, and therefore it cannot be limited to the theoretical model from which simulations originate. Morrison is right in persuading us that the representational capacity of the simulation model can also be approached during the modeling stages. And Lenhard is right in pointing out the relative autonomy of computer simulations from models, data and phenomena.

Whereas it is true that some researchers use for their simulations 'off-the-shelf' computer software, it is not uncommon to find many other researchers designing and programing their own simulations. The reason is that there is more than an intellectual satisfaction in programming your own simulations, there are in fact good epistemic reasons for doing so. By getting involved in the design and programing of computer simulations, researchers get to know what is being designed and implemented, and

---

[13] This can be found in Oberkampf et al. (2003).

[14] See, for instance, his work in Winsberg (2010).

[15] For more on these ideas, see Morrison (2009).

therefore, they have a better idea of what should be expected in terms of errors, uncertainties, and the like. In fact, the more involved researchers are in the design, programing, and implementation of computer simulations, the better prepared they are for understanding their simulations.

There are two general sources that feed the methodology of computer simulations. On the one hand, there are *formal rules and systematic methods* that allow formal equivalence between mathematical models and simulation models. This is illustrated by some discretization methods that allow the transformation of continuous functions into its discrete counterparts. Examples of discretization methods are the Runge-Kutta method and the Euler method.[16] There are also many formal development languages targeted at the analysis of designs and identification of key features of modeling, including errors and misrepresentations. Such languages, like Z notation, VDM, or LePUS3, typically use some formal programing language semantics (i.e., denotational semantics, operational semantics, and axiomatic semantics) for the development of computer simulations and computer software in general.

The second source for a methodology of computer simulations has a more practical side. The design and programing of computer simulations do not depend solely on formal machinery, but also rely on the so-called *expert knowledge* (Collins and Evans 2007). Such knowledge includes 'tricks', 'know-how', 'past experiences,' and a host of non-formal mechanisms for designing and programing their simulations. It is not easy to pin down the form and uses of such knowledge as it depends on many factors, such as communities, institutions, and personal educational background. Shari Lawrence Pfleeger and Joanne M. Atlee report that expert knowledge, as present and as necessary as it is in the life of computer software, is nevertheless subjective and dependent on the current information available, like accessible data and the actual development state of the software unit.[17] In this context, it is not surprising that many decisions about computer software that were traditionally made by the expert researcher are now being carried out by automated algorithms.

A synthesis of these two sources consists in relying on well-documented software packages. On the one hand, many of these packages have been formally checked; on the other, the teams involved in their development are the experts on the subject. A case in point are *random number generators*, which are at the heart of stochastic simulations such as Monte Carlo simulations. For these software packages, it is important to formally show that they behave in the way they were specified, for in this way researchers know the limits of its functionality—lower and upper pseudo-random number generated—possible errors, repetitions, and the like. Choosing the right random number generation engine is an important design decision, as accuracy and precision of the results depend on it.[18]

The two sources mentioned above are present during the many stages of the design and programing of computer simulations. To neglect non-formal routines, for

---

[16]For a discussion on discretization techniques, see Atkinson et al. (2009), Gould et al. (2007), and Butcher (2008) among others in a very large body of literature.

[17]For more on the life cycle of computer software, see (Pfleeger and Atlee 2010).

[18]The loci classici on this topic are Knuth (1973) and Press et al. (2007).

instance, leads to the wrong idea that there is an axiomatic method for designing computer simulations. Likewise, failing to acknowledge the presence of systematic and formal methods leads to a view of computer simulations as an unstructured, ungrounded discipline. The challenge, then, is to understand the role that each one plays in the complex activity that is the methodology of computer simulations. It is also true that, despite the many ways in which computer simulations are designed and programed, there is a standardized methodology that provides the basic grounds for many computer simulations. It is hard to imagine a team of scientists and engineers changing their general methodologies every time they design and program a new computer simulation. The methodology for computer simulations, we could say, relies on ideals of stability of behavior, reliability, robustness, and continuity in design and programing.[19]

In the following, I discuss a methodology for computer software in general, and for computer simulations in particular.[20] The idea is to have a general overview of what comprises a typical simulation model, and what it means to be a computer simulation. I shall also address some issues that caught the attention of philosophers. I hope that by the end of this section we are convinced that computer simulations represent a new unit of analysis for science and engineering.

2.2.1.1 Specifications[21]

Every scientific instrument requires the elucidation of its functionality and oper-ability, design and constraints. The kind of information here laid out constitutes the *specification* of that scientific instrument. Take for instance the mercury-in-glass thermometer with the following specification:

1. Insert mercury in a glass bulb attached to a glass tube of narrow diameter; the volume of the mercury in the tube must be much less than the volume in the bulb; calibrate marks on the tube that vary according to the heat given; fill the space above the mercury with nitrogen.
2. This thermometer can only be used for measuring liquids, body temperature, and weather; it cannot measure above 50 °C, or below −10 °C; it has an inter-

---

[19] A good example of this, but at the level of computer hardware, is von Neumann's architecture, which has been a standard in the design of computers since his 1945 report (Von Neumann 1945). Naturally, one needs to also take into consideration the modifications made in the computer archi-tecture brought up by technological change.

[20] Let me note that while I present only three constituents of computer simulations, the possibility of having up to six has been argued. Giuseppe Primiero recognizes up to six 'levels of abstractions' at which computational systems are examined, namely, Intention, Specification, Algorithm, High-level programing language instructions, Assembly/machine code operations, an execution (Primiero 2016). There are many philosophical approaches to the nature of specifications, algorithms, and computer processes that I will not be able to discuss. A short-list of references would certainly include Copeland (1996); Piccinini (2007, 2008); Primiero (2014) and Zenil (2014).

[21] The following discussion can also be found in Spanish in (Durán 2018b). Published with the permission of Principia: An International Journal of Epistemology.

val of 0.5 °C between values and an accuracy of ±0.01 °C; the mercury in the thermometer solidifies at −38.83 °C;

3. Instructions for the correct use of the thermometer: insert the mercury bulb located at the end of the thermometer into the liquid (under the arm, or outside but not in direct sunlight) to be measured, locate the value indicated by the height of the mercury bar, and so forth.

In addition to these specifications, we typically know relevant information about the target system. In the case of the thermometer, it might be relevant to know that water changes state from liquid to a solid at 0 °C; that if the liquid is not properly isolated, then the measurement may be biased by a different source of temperature; that the Zeroth law of thermodynamics justifies the measurement of the physical property 'temperature,' and so forth.

Along with laying out the necessary information for building an instrument, the specification is also fundamental for establishing the reliability of the instrument and correctness of its results. Any misuse of the thermometer, that is, any use that explicitly violates its specifications, might lead to inaccurate measurements. Conversely, to say that a thermometer carries out the required measurement, that the measurement is precise and accurate, and that the values obtained are reliable measurements, is also to say that the measurement has been carried out within the specifications given by the manufacturer.

An anecdotal case of misuse of an instrument has been brought up by Richard Feynman during his time on the research committee investigating the Challenger disaster. He recalls having the following conversation with the manufacturer of an infrared scanning gun:

> Sir, your scanning gun has nothing to do with the accident. It was used by the people here in a way that's contrary to the procedures in your instruction manual, and I'm trying to figure out if we can reproduce the error and determine what the temperatures really were that morning. To do this, I need to know more about your instrument. (Feynman 2001, 165–166)

The situation was probably very frightening for the manufacturer, as he might have thought that the scanning gun did not work as specified. That was in fact not the case. Instead, it was the material used in the shuttle's O-rings that became less resilient in cold weather, and therefore not properly sealing on an unusually cold day at Cape Canaveral. The point here is that to Feynman, as well as to any other researcher, the specification is a key piece of information about the proper design and use of an instrument.

Thus understood, specifications fulfill a *methodological* purpose as well an *epistemic* functionality. Methodologically speaking, it works as the 'blueprints' for the design, construction, and use of an instrument. Epistemically speaking, it works as the depository and repository of our knowledge about that instrument, its potential outcomes, errors, etc. In this context, the specification has a double aim. It provides relevant information for the construction of an instrument as well as insight into its functionality. From the example of the thermometer above, point 1 illustrates how to construct a thermometer, including how to calibrate it; point 2 illustrates the upper and lower boundaries in which the thermometer measures and can be used as a reliable instrument; point 3 illustrates the correct use of the thermometer for successful measurements.

In the context of computer simulations,[22] specifications play a similar role as with instruments: they are descriptions of the behavior, components, and capabilities of the computer simulation in accordance with the target system. Brian Cantwell-Smith, a philosopher interested in the foundations of computing, defines it as a "formal description in some standard formal language, specified in terms of the model, in which the desired behavior is described" (Cantwell 1985, 20). The example used is of a milk-delivery system, which can be minimally specified as a milk delivery car that visits each store driving the shortest possible distance in total.

Let us stop here for a moment and analyze Cantwell-Smith's definition. Although illuminating, this definition does not capture what researchers call today a 'specification'. There are two reasons for this. First, it is reductive. Second, it is not sufficiently inclusive. It is reductive because the notion of specification is identified with a *formal description* of the behavior of the target system. This means that the intended behavior of the simulation is described in terms of formal machinery. Software engineering, however, has made it clear that specifications cannot be fully formalized. Rather, they must be conceived as some form of 'semi-formal' descriptions of the behavior of a computer software. In this latter sense, formal as well as non-formal descriptions coexist in the specification. In other words, mathematical and logical formulae coexist with design decisions in plain natural language, *ad-hoc* solutions to computational tractability, and the like. Along comes the documentation for the computer code, the researchers' comments on it, and so on. Although there is general agreement that full formalization of the specification is a desirable aim, it is not always an attainable one, especially for the kind of very complex specifications that computer simulations presuppose.

The definition is not sufficiently inclusive in the sense that specifying the proper behavior of a simulation is *not* independent of how it is implemented. A specification does not only describe the intended behavior of the simulation in accordance to the target system, but also incorporates practical and theoretical constraints of the physical computer. This means that worries about computability, performance, efficiency, and robustenness of the physical computer are usually part of the specification of the simulation as well. In the above example of the milk-delivery simulation, the specification only captures the general aspects of the target system, those that are more important for the researcher regardless of details about how the delivery of milk is actually carried out. This much is correct. On top of this, the constraints tailored to the physicality of the computer need to be added. As a result, the specification will be populated with new entities and relations, alternative solutions to the problems at hand, and an overall different approach than a purely formal solution would provide.

---

[22]Let it be note that the differences between the specification for a computer simulation and any other computer software are minimal. This is due the fact that computer simulations are a kind of computer software. In this respect, the most concrete difference lies on the kind of model implemented, namely, a model that describes a target system intended for general scientific and engineering purposes.

A more precise idea of what constitutes specifications is given by Shari Pfleeger and Joanne Atlee.[23] These mathematicians and computer scientists believe that any program specification amounts to describing external visible properties of a software unit, along with the system's access functions, parameters, return values, and exceptions. In other words, the specification takes care of both, the modeled target system as well as the software unit.

There are a series of general attributes and features that are typically included as part of the specification. The following is a shortlist:

Purpose: it documents the functionality of each access function, modification of variables, access to I/O, etc. This should be done in enough detail that other developers can identify which functionalities fit their needs;

Preconditions: these are assumptions that the model includes and which must to be available for other developers to know under what conditions the unit works correctly. Preconditions include values of input parameters, states of global resources, other software units, etc.;

Protocols: this includes information about the order in which access functions should be invoked, the mechanisms in which modules exchange messages, etc. For example, a module accessing an external database needs to be properly authorized;

Postconditions: all possible visible effects of accessing functions are documented, including return values, exceptions, output files, etc. Postconditions are important because they tell the calling code how to react appropriately to a given function's output;

Quality attributes: these are the performance and reliability of the model visible to developers and users. In the example of the orbiting satellite around a planet in Sect. 1.1, the user is required to specify the TOLERANCE variable, that is, the maximum absolute error that can be tolerated in any positional coordinate. If this is set too low, the program can become very slow;

Design decisions: the specification is also the place where political, ethical, and design decisions are implemented as part of the simulation model.

Error treatment: it is important to specify the working flow of the software unit and how to behave in abnormal situations, such as invalid input, errors during the computation, handling errors, and so on. The functional requirements of the specification must clearly state what the simulation must do in these situations. A good specification must follow specific principles of robustness, correctness, completeness, stability, and similar desiderata.

Documentation: any extra information must also be documented into the specification, such as specifics of the programming languages used, libraries, relations held by structures, extra-functions implemented and so forth;

In this vein, and similarly to scientific instruments, specifications play two central roles: they play a *methodological* role as the blueprint for the design, programing, and

---

[23] See Pfleeger and Atlee (2010).

implementation of the simulation. Included in this role is linking the representation and knowledge about the target system together with knowledge about the computer system (i.e., the computer's architecture, OS, programming languages, etc.). This means that specifications help 'bridge' connecting these two kinds of knowledge. This is of course not a spooky or mysterious connection, but rather is the basis of the professional activity of computer scientists and engineers: a simulation must be specified in as much detail as possible before it is programmed into an algorithm, for it saves time, money, resources and, more importantly, reduces the presence of errors, misrepresentations, and the chances of miscalculations.

It also plays an *epistemological* role in the sense that specifications are a depository and repository of our knowledge about the target system and, as just mentioned, of the software unit as well. In this sense, specifications are a cognitively transparent unit in the sense that researchers can always understand what has been described there. In fact, it can be argued that it is the most transparent unit in a computer simulation. This is immediately clear when compared with the algorithm and the computer process: the former, although still cognitively accessible, is written in some programming language unsuitable for a human to follow throughout; the latter, on the other hand, obscures any access to the computer simulation and its development over time.

Let us illustrate these points with an example. Consider the specification of a simple simulation, such as the simulation of the orbiting satellite under tidal stress discussed in Sect. 1.1. A possible specification includes information about the behavior of the satellite, such as the fact that it stretches along the radius vector. It also includes possible constraints to the simulation. Woolfson and Pert indicate that if the orbit is non-circular, then the stress on the satellite is variable and therefore it expands and contracts along the radius vector in a periodic fashion. Because of this, and to the effect that the satellite is not designed to be perfectly elastic, there will be hysteresis effects and some mechanical energy will be converted into heat which is radiated away. An interesting *ad-hoc* solution to the elasticity of the satellite consists in representing it by three masses, each of the same value connected by springs of the same unstrained length. Naturally, a number of equations are also included in the specification as well.[24]

These and other elements must then be included into the specification along with some information about the physical computer. A simple example of this last point is that the mass of Jupiter cannot be represented on a computer whose architecture is based on a 32 bits system. The reason is because the mass of Jupiter is 1.898×10^{27} kg, and such number can only be represented in 128 bits or higher.

Let us now continue with the study of the algorithm, that is, the logical structure in charge of interpreting the specification in a suitable programming language.

---

[24]For more philosophical questions about specifications, see Turner (2011).

2.2.1.2 Algorithms

Most of our daily activities can be described as simple sets of rules that we systematically repeat. We wake up at a certain time in the morning, brush our teeth, take a
shower, and take the bus to work. We modify our routine, of course, but just enough
to make it more advantageous in some way: it gives us more time in bed, it minimizes
the distance between stops, it satisfies everyone in the house.

In a way, this routine captures the sense of what we call an *algorithm*, as it
systematically repeats a set of well defined rules over and over. Jean-Luc Chabert, an
historian that dedicated much time to the notion of algorithm, defines it as "a set of
step by step instructions, to be carried out quite mechanically, so as to achieve some
desired result" (Chabert 1994, 1). A routine is a kind of algorithm or, to be more
precise, it could be turned into an algorithm. But in itself it is not an algorithm. We
need, then, to make this notion more precise.

Let us first acknowledge that the notion of an algorithm existed well before a word
was coined to describe it. In fact, algorithms have a pervading history that goes back
to the Babylonians and their use for decodings points of Law, Latin teachers who
used algorithms to elaborate on grammar, and clairvoyants for predicting the future.
Their popularity begins with mathematics—Euler's and Runge-Kutta's methods, and
the Fourier series are just a few examples—and spreads into computer science and
engineering. This last stop is where our interest in algorithms lies. In short, to us,
algorithm is synonymous with *computer algorithm*.[25]

Now, computer algorithms rest on the idea that they are part of a systematic,
formal, and finite procedure—mathematical and logical – for implementing specific
sets of instructions. Chabert, borrowing from the Ecyclopaedia Britannica, defines
them as "a systematic mathematical procedure that produces—in a finite number of
steps—the answer to a question or the solution of a problem" (Chabert 1994, 2).
Under this interpretation, then, the following collision-less particle-in-cell system as
presented by Michael Woolfson and Geoffrey Pert qualifies as an algorithm:

1. Construct a convenient grid in the one-, two-or three-dimensional space within
   which the system can be defined. [...]
2. Decide on the number of superparticles, both electrons and ions, and allocate
   positions to them. To obtain as little random fluctuation in the fields as possible
   it is required to have as many particles as possible per cell. [...]
3. Using the densities at grid-points, Poisson's equation: $\nabla^2 \phi = -\rho/\varepsilon_0$

   ...

N. If the total simulation time is not exceeded then return to *3*. [26]

---

[25]There is an interesting etymological origin for the word 'algorithm'. Records show that the word
partly derives from al-Khwārizmī, a ninth century Persian mathematician, author of the oldest
known work of algebra. The world also comes from the Latin *algorismus* and the ancient Greek
$\alpha\rho\iota\theta\mu\acute{o}\varsigma$, meaning 'number'.

[26]A more detailed discussion of the steps involved in the process for purely electrostatic fields can
be found here (Woolfson and Pert 1999a, 115).

A close look reveals that the example includes mathematical machinery as well as statements in plain English. In a way, it looks more like a specification for a collision-less particle-in-cell system than a proper computer algorithm. And yet, it qualifies as an algorithm under Chabert's definition. What we need is a more precise definition.

In the 1930s, the concept of computer algorithm was popularized in the context of computer programming. In this new context, the notion suffered some alterations from its original mathematical formulation, most prominently in the based language: from mathematics to a manifold of syntactic and semantic constructions. However, that was not all. A shortlist with the new features include:

1. An algorithm is defined as a finite and organized set of instructions, intended to provide the solution to a problem, and which must satisfy certain sets of conditions;
2. The algorithm must be capable of being written in a certain language;
3. The algorithm is a procedure which is carried out step by step;
4. The action at each step is strictly determined by the algorithm, the entry data and the results obtained at previous steps;
5. Whatever the entry data, the execution of the algorithm will terminate after a finite number of steps;
6. The behavior of the algorithm is physically instantiated during the implementation on the computer machine.[27]

Let it be noticed that many structures in computer science and engineering equally succeed in qualifying as algorithms with these features. *Pseudo-codes* (e.g., Algorithm 1) is our first example. These are descriptions that fulfill most of the above conditions, except that they are not implementable on the physical machine. The reason is that they are primarily intended for human reading, with a dash of formal syntax. Because of this, pseudo-code is quite frequently found in the specifications of a target system, as they facilitate the transition into an algorithm.[28]

Most commonly, the notion of algorithm is tailored to a *programming language*. Now, since the universe of programming languages is significantly large, one example of each kind would suffice for our purposes. I then take Fortran, Java, Python, and Haskell as four representatives of programming languages. The first is an example of imperative programming language (see Algorithm 2); the second is an Object Oriented Programming Language (see Algorithm 3); Python is a good example of an interpreted language (see Algorithm 4); and Haskell is the exemplar of functional programming (see Algorithm 5).

Algorithms show several characteristics worth paying attention to. From an ontological point of view, an algorithm is a *syntactic structure* encoding the information

---

[27]These are some, but not all of the characteristics ascribed to algorithms by Chabert. A more detailed history of computer algorithms can be found in (Chabert 1994, 455).

[28]For a thorough discussion on the different notions of algorithm, see the debate between Hill and Robin (2013; 2016) and Andreas Blass, Nachum Dershowitz and Yuri Gurevich in (Blass and Gurevich 2003; Blass et al. 2009). Here we do need to delve deeply into subtle philosophical discussions.

Algorithm 1 Pseudo-code

Pseudocode is a non-formal, high-level description of the specification. It is intended to focus on the operational behavior of the algorithm rather than on a particular syntax. In this sense, it uses a similar language as a programming language but in a very loose sense, typically omitting details that are not essential for understanding the algorithm.

Require: $n \geq 0 \vee x \neq 0$
Ensure: $y = x^n$
  $y \leftarrow 1$
  if $n < 0$ **then**
    $X \leftarrow 1/x$
    $N \leftarrow -n$
  else
    $X \leftarrow x$
    $N \leftarrow n$
  end if
  while $N \neq 0$ **do**
    if N is even **then**
      $X \leftarrow X \times X$
      $N \leftarrow N/2$
    else $\{N$ is odd$\}$
      $y \leftarrow y \times X$
      $N \leftarrow N - 1$
    end if
  end while

---

set out in the specification. As I will discuss later in this section, for several technical and practical reasons, algorithms cannot encode all the information set out in the specification. Rather, some information will be added, some will be lost, and some will simply be altered.

Let us also note that studies in computer science make use of the notion of *syntax* and *semantics* in a rather different way than linguistics. For computer science, *syntax* is the study of symbols and their relationships within a formal system; it typically includes a grammar (i.e., a sequence of symbols as well-formed formulas), and proof-theory (i.e., a sequences of well-formed formulas that are considered theorems). On the other hand, *semantics* is the study of the relationship between a formal system, which is syntactically specified, and a semantic domain, which is specified by a domain providing interpretation to the symbols in the syntactic domain. In the case of the implementation of the algorithm on the digital computer, the semantic domain is the physical states of the computer when running the instructions programed into the algorithm. I refer to these physical states as the *computer process*, and it will be discussed in the next section.

Two other ontological properties of interest are that the algorithm is *abstract* and *formal*. It is *abstract* because it consists of a string of symbols with no physical causal relations acting upon them: just like a logico-mathematical structure, an algorithm

---

Algorithm 2 Fortran

Fortran is an imperative programming language that is especially suited to numeric computation and scientific computing. It is quite popular among researchers working with computer simulations, as formulas are easily implemented while computing performance remains high. Due to its versatility and efficiency, Fortran came to dominate in computationally intensive areas such as numerical weather prediction, finite element analysis, and computational fluid dynamics, among others. It is also a very popular language in High-Performance Computing.

Program *GDC* calculates the Greatest Common Divisor between two integers as input by the user:

```
program GCD
        implicit none
        integer :: a, b, c

        write(*,*) 'Give me two positive integers: '
        read(*,*) a, b
                if (a < b) then
                c = a
                a = b
                b = c
        end if

        do
                c = MOD(a, b)
                if (c == 0) exit
                a = b
                b = c
        end do
        write(*,*) 'The GCD is ', b
end program GCD
```

---

is causally inert and disconnected from space-time. It is *formal* because it follows the laws of logic that indicate how to systematically manipulate symbols.[29]

These ontological features offer several epistemic advantages of interest, two of which have kept philosophers and computer scientists particularly busy. Those are *syntax correlation* and *syntax transference*. Syntax correlation is the possibility of upholding equivalence between two algorithmic structures. Syntax transference, on the other hand, consists in altering the algorithm to make it perform a different functionality. Let me elaborate more on each.

Syntax correlation can be made clear with an example from mathematics. Consider a cartesian system of coordinates and a polar system of coordinates. There

---

[29]For more details, see the work of Donald E. Knuth and Edsger W. Dijkstra in Knuth (1973, 1974) and Dijkstra (1974). Both authors are recognized as major contributors in the development of computer science as a rigorous discipline. Knuth is also the author of *The Art of Computer Programming*, a four volume masterpiece—by today's count—that covers many kinds of programming algorithms and their analysis. Dijkstra, in turn, is an early pioneer and promoter of computer science as an academic discipline. He helped lay the foundations for the birth and development of software engineering, and his writing set the baseline in many research areas of computing science, especially in structured programming and concurrent computing.

Algorithm 3 JAVA

Java is a general-purpose computer programming language. Some of its main characteristics are that it is concurrent, class-based, object-oriented, and specifically designed to have as few implementation dependencies as possible. This last point has made Java a popular programming language since its applications are typically compiled once and run on any Java Virtual Machine, regardless of the computer architecture.

SNP calculates the sum of two integers as input by the user:

```
package SNP;
import java.util.Scanner;

public class addTwoNumbers
        private static Scanner sc;
        public static void main(String[] args)
                int Number1, Number2, Sum;
                sc = new Scanner(System.in);

                System.out.println("Enter the first number: ");
                a = sc.nextInt();

                System.out.println("Enter the second number: ");
                b = sc.nextInt();

                sum = a + b;
                System.out.println("The sum is = " + sum);
```

Algorithm 4 Python

Python is a widely used high-level programming language used for general-purpose programming. It is one of the many interpreted languages available today. One of Python's design philosophies is code readability, enforced by using whitespace indentation to delimit code blocks. It also makes use of a remarkably simple syntax that allows programmers to express complex concepts in fewer lines of code.

Function *gcd* calculates the Greatest Common Divisor of two integers as input by the user:

```
def gcd(a, b):
        a = abs(a)
        b = abs(b)
                while b:
                        a, b = b, a%b
                return a
```

Algorithm 5 Haskell

Haskell is a standardized, general-purpose, purely functional programming language, with non-strict semantics and strong static typing. Being a functional language means that Haskell treats computation as the evaluation of mathematical functions. In this respect, programming in Haskell is done with expressions or declarations instead of statements—as opposed to imperative programming. Haskell features lazy evaluation, pattern matching, list comprehension, type classes, and type polymorphism. One main characteristic of functions in Haskell is that they have no side effects, that is, the result of a function is determined by its input and only by its input. Functions, then, can be evaluated in any order and will always return the same result—provided that the same input is passed. This is a major advantage of functional programming that makes the behavior of a program much easier to understand and to predict.

Function *plus* adds 1 to 2 and shows the result:

plus :: Int − > Int − > Int
plus = (+)

main = **do**
      let res = plus 1 2
      putStrLn $ "1+2 = " ++ **show** res

---

are well established mathematical transformations that help set their equivalence. Given cartesian coordinates (x, y), their equivalent polar coordinates is set by $(r, \theta) = (\sqrt{x^2 + y^2}, \tan^{-1} \frac{y}{x})$. Likewise, given a set of polar coordinates (r, θ), one can find the corresponding cartesian coordinates without much effort using $(x, y) = r \cos\theta, r \sin\theta$).[30]

A similar idea can be used in algorithms. Consider the following extraction of Algorithm 6 and its equivalent in Algorithm 7. We know that both are logically equivalent because there is a formal proof of this (given in the truth Table 2.1).[31]

Algorithm 6

...**if** (α) **then** $\{a\}$ **else** $\{b\}$ ...

---

[30] As it turns out from the example, syntax correlation raises the question 'to what extent are two systems of equations equivalent?' This is not an easy question to answer. In the case of cartesian and polar systems of coordinates, one could object that restrictions must be imposed on the polar coordinates systems—i.e., for the *tan* function, the domain is all real numbers except $\pm\frac{\pi}{2}, \pm\frac{3\pi}{2}, \pm\frac{5\pi}{2}, ...$, where the function is undefined—and therefore they are not *isomorphic*. Something similar happens with Lagrangian and Hamiltonian, as I discuss below for algorithms.

[31] Surely, algorithms are much more complex structures than the examples I use here. It comes as no surprise, then, that the equivalence between algorithm cannot be simply established by a truth-table. More complex mathematical and computational machinery is needed and, indeed, used. One standard approach consists in constructing an equivalence class of algorithms (Blass et al. 2009). Take for instance a sorting algorithm, whose formulation is that it returns an ordered permutation of an input list, for some definition of ordering. By showing that a given function has the property of returning an ordered permutation—for the same definition of ordering—then one could assert that both algorithms belong to the same class. Finding an algorithm equivalence is central for many software and hardware verification procedures.

Table 2.1 Equivalence truth-tables of Algorithms 6 and 7

α	$\{a\}$	$\{b\}$
T	T	$\emptyset$
F	$\emptyset$	T

not-α	$\{a\}$	$\{b\}$
T	$\emptyset$	T
F	T	$\emptyset$

Algorithm 7

... **if** (**not-α**) **then** $\{b\}$ **else** $\{a\}$ ...

One epistemic advantage of syntax correlation is that it expands the number of possible and equivalent implementations for any given computer simulation. Consider as a case a Lagrangian and a Hamiltonian set of equations, as they are correlated in dynamic systems. Researchers get to choose between one or the other formulation depending on needs that are not strictly related to the representation of the target system (e.g., understanding one system of equations is simpler than the other for a given target system, the performance of the simulation is improved with either set of equations, etc.). In this way, researchers are no longer stuck with one set of equations figuring out how to implement them, but rather they focus efforts and concerns on other aspects of the simulation such as performance and simplicity. Take for instance the following quick calculation. For a system with configuration space of dimension n, Hamiltonian equations are a set of $2n$ coupled first-order ODEs. Lagrangian equations, on the other hand, are a set of n uncoupled, second-order ODEs. Thus, implementing Hamiltonians over Langrangian could give a real advantage in terms of performance, memory use, and speed of calculations. Something very similar happens when one calculates a cartesian and a polar systems of coordinates by hand.

The second epistemic feature of algorithms is *syntax transference*. This refers to the simple idea that by adding—or subtracting—just a few lines of code in the algorithm, researchers are able to reuse the same code for different representational contexts. In such cases, syntax transference presupposes minimal changes in the algorithm. A very simple case can be illustrated by adding to Algorithm 3 a few lines for computing the summation of 3 numbers, as shown in Algorithm 8. Let us note that syntax transference is a foundational idea behind modules and libraries: a very similar code can be used in different, but related, contexts.

Syntax transference, then, allows researchers to reuse their existing code in order to accommodate it to different contexts, as well as to generalize the code in order to include more results, thus broadening—or narrowing—the scope of the algorithm.

Syntax correlation and syntax transference are common practices among researchers programming their own simulations. It is not unusual to see how the same simulation grows and shrinks by adding and eliminating some modules as well as modifying some others. This is part of standard code maintenance and improvement. Now, it is also possible that syntax transference makes the code grow too bulky, and thus impossible to maintain without a height cost. When such a situa-

---

Algorithm 8 JAVA expanded

```java
package SNP;
import java.util.Scanner;

public class addThreeNumbers
        private static Scanner sc;
        public static void main(String[] args)
                int Number1, Number2, Number3, Sum;
                sc = new Scanner(System.in);

                System.out.println("Enter the first number: ");
                a = sc.nextInt();

                System.out.println("Enter the second number: ");
                b = sc.nextInt();

                System.out.println("Enter the third number: ");
                c = sc.nextInt();

                sum = a + b + c;
                System.out.println("The sum is = " + sum);
```

---

tion occurs, it is probably time for a new code. For cases like this, syntax correlation plays an important role, as many functions of the old code will be kept—and properly modified—in the new code.

Attached to syntax correlation and syntax transference come some philosophical issues. The most prominent one is the question 'when are two algorithms the same?' The question emerges in the context that syntax correlation and syntax transference presuppose modifications of an original algorithm, leading to a new algorithm. In the case of syntax correlation, this comes in the form of a new algorithm performing the same functionalities as the old algorithm. In the case of syntax transference, this comes in the form of having a modified—and thus, strictly speaking, a new—algorithm.

There are two generally accepted answers to this question. Either two algorithms are *logically equivalent*, that is, the two algorithms are structurally and formally similar,[32] or they are *behaviorally equivalent*, that is, the two algorithms behave in a similar fashion. Let me briefly discuss these two approaches.

Logical equivalence is the idea that two algorithms are structurally similar, and that such equivalence can be shown by formal means. I illustrate a very simple logical equivalence by using Algorithms 6 and 7, since both are formally isomorphic to each other—the proof, again, is in Table 2.1. Formal procedures of any kind—like a truth

---

[32]Isomorphism would be the best option here, since it is the only–morphism that could warrant total equivalence between algorithms. Alternative–morphisms, however, are also discussed in the literature, for instance, in the work of Blass et al. (2009) and Blass and Gurevich (2003).

table—are good warrants for structural similarity.[33] Thus, the *if ...then* conditional in algorithm 6 is structurally equivalent to the conditional in Algorithm 7.

Unfortunately, logical equivalence is not always attainable due to practical as well as theoretical constraints. Examples of practical constraints include cases of algorithms that are humanly impossible to formally verify. Another example is formal procedures which are too time and resources consuming. Examples of theoretical constraints include cases where the programming language coded in the simulation refers to entities, relations, operations, and the like, which the procedure for checking structural similarity is unable to account for.

To deal with these constraints, academia and industry have joined forces and created a host of tools that automate the process of proofing and checking. Although there are several model checkers and semantics available, one powerful example used in verification of software is ACL2. A Computational Logic for Applicative Common Lisp (ACL2) has been explicitly designed to support automated reasoning and thus help in the reconstruction of equivalence classes of algorithms.

Behavioral equivalence, on the other hand, consists in making sure that the two algorithms behave in a similar manner (e.g., by producing the same results[34]). Now, although behavioral equivalence sounds simpler to achieve than structural equivalence, it carries some problems of its own. For instance, there is the concern that behavioral equivalence is grounded on inductive principles. This means that one could only warrant equivalence up to time t, when the two algorithms behave similarly. But there are no warrants that at time $t + 1$ the algorithms' behavior will still be the same. Behavioral equivalence can only be warranted up to the time the two algorithms are being compared. In fact, one could make the case that two algorithms are behaviorally equivalent only for *that* execution of the algorithms. Further executions could show divergence in behavior.[35]

A final issue stemming from behavioral equivalence is that it could hide logical equivalence. This means that two algorithms diverge in behavior although they are structurally equivalent. An example is an algorithm that implements a cartesian set of coordinates whereas another implements polar coordinates. Both algorithms are

---

[33] For those readers interested in deep philosophical discussions, clarifying the notion of 'structural similarity' is a must. Here, I take it that one can objectively decide when one algorithm is structurally similar to another. The literature on similarity and, more generally, on theoretical representation is quite vast. A suggestion is to begin with Humphreys and Imbert (2012).

[34] This under a given interpretation of 'the same results,' otherwise we are begging the question of when two sets of results are the same. To this end, we could select mathematical and algorithmic procedures external to the two algorithms being compared that establish acceptable similarity among results.

[35] Many researchers are reluctant to accept as a genuine concern that computer code could behave differently depending on the time of execution. The philosopher James Fetzer Fetzer (1988) brought up this issue once in the context of program validation. In response, many computer scientists and engineers objected that he knew little of how computer software and hardware actually work. Of course, the objections were not merely *ad hominem*, but contained good reasons to reject Fetzer's position. In any case, Fetzer raised a genuine philosophical issue and it should be treated as such.

structurally equivalent, but behaviorally dissimilar.[36] In these kind of cases, it remains the question of the kind of equivalence that should prevail.

These are some of the standard discussions found in the philosophy of computer science. In here, I have only scratched the surface, and much more can and needs to be said. The lesson I would like to draw, however, is that both syntax correlation and syntax transference come at a price. Whether researchers are willing to pay that price—and how high that price actually is—is a question that depends on several variables, such as the researcher's interests, the available resources, and the real urgency of finding a solution. For some situations, ready out-of-the-box solutions exist; for some others, the researcher's experience is still the most valuable currency.

Until now, we have discussed algorithms along with their philosophical consequences. We still need to say something about the link between the specification and the algorithm.

Ideally, the specification and the algorithm should be closely related, that is, the specification should be completely interpreted as an algorithmic structure. In reality, this is rarely the case mostly because the specification makes heavy use of natural language while in the algorithm every term needs to be interpreted literally. A case in point is the use of metaphors and analogies. Many scientific and engineering models include terms that have no specific interpretation, like 'black hole' or 'mechanism' (Baller-Jones 2009).[37] Metaphors and analogies are then used to fill the gap introduced by these terms which have no literal interpretation. By doing so, metaphors and analogies inspire some kind of creative response in the users of the model that cannot be rivaled by literal language. However, if the same metaphorical terms are implemented in an algorithm, they require a literal interpretation otherwise they cannot be computed.

Of course, to make the interpretation of the specification into an algorithm easier, researchers have once more relied on the automatization provided by computers. To this end, there is a host of specialized languages that formalize the specification, facilitating in this way its programming into an algorithm. Common Algebraic Specification Language (CASL), Vienna Development Method (VDM), Abstract Behavioral Specification (ABS), and the Z notation, are just a few examples.[38] Model checking is also useful since it automatically tests whether an algorithm meets the required specification, and therefore it also helps with its interpretation. In short, the interpretation of a specification into an algorithm has a long tradition in mathematics, logic, and computer science, and it does not really represent a conceptual problem here.

I also mentioned that the specification includes non-formal elements, such as expert knowledge and design decisions that cannot be formally interpreted. However, these non-formal elements must also be included—and included correctly—into the

---

[36]Let us note that this reasoning depends on the notion of 'behavior.' If by this we simply take 'exactly the same results,' the two algorithms clearly render diverse results.

[37]We have to be cautious here because, in cases like neuroscience, terms such as 'mechanism' have a full fledged definition (e.g., Machamer et al. 2000, and Craver 2001).

[38]For CASL, see for instance Bidoit and Mosses (2004). For VDM, see for instance Bjorner and Henson (2007). For ABS, see http://abs-models.org/concept/. And for the Z notation, see for instance Spivey (2001).

algorithm, otherwise they will not be part of the computation of the model. Imagine for instance the specification for a simulation of a voting system. For this simulation to be successful, statistical modules are implemented in such a way that will give a reasonable distribution of the voting population. During the specification stage, researchers decide to give more statistical relevance to variables such as sex, gender, and health over other variables, like education and income. If this design decision is not programed into the statistical module appropriately, then the simulation will never reflect the value of these variables, despite being in the specification.

This example is meant to show that an algorithm must be capable of interpreting formal as well as non-formal elements included in the specification, particularly because there are no formal methods for interpreting expertise knowledge, past experiences, and the like.

Where does this discussion leave us in the methodological map of computer simulations? On the one hand, we have a better grasp of the nature of computer simulations as units of analysis. On the other hand, we have a deeper understanding of the methodology of specifications, algorithms, and their relation.

Before continuing, let me clarify some of the terminology that I have been using. Call *computer model* the whole process of specifying and programing a given computer system. If the purpose of such model is for simulating a target system, then let us call it the *simulation model*. The only visible differences between these two is that the former is a generalized version of the latter. More significant differences will emerge when discussing epistemological and practical matters in the following chapters. In addition, by implementing a computer model on the physical computer we obtain a *computer process* (to be discussed next). Following the same structure as before, if the model implemented is a simulation model, then we have a *computer simulation*.

2.2.1.3   Computer Processes

In previous sections, I made use of Cantwell-Smith's notion of specification as the starting point for our studies. Now it is time to complete his idea with the analysis of computer processes.[39]

Following the author, a *computer program* is the set of instructions taking place on the physical computer. Such a characterization differs greatly from the notion of specification, as discussed earlier by Cantwell-Smith. While a computer program is a causally-related process taking place on the physical computer, the specification is an abstract—and, to some extent, formal—entity. In addition, Cantwell Smith points out that "the program has to say *how the behavior is to be achieved*, typically in a step by step fashion (and often in excruciating detail). The specification, however, is less

---

[39] I do not address the question about the computational architecture on which the computer software is run. However it should be clear from the context that we are interested in silicon-based computers—as opposed to quantum computers, or biological computers. It must also be said that other architectures represent a different set of problems. (see, for instance, Berekovic et al. (2008) and Rojas and Hashagen (2000)).

constrained: all it has to do is to specify *what proper behavior would be*, independent of how it is accomplished" (Cantwell 1985, 22; emphasis in the original.). Thus understood, specifications are *declarative* in the sense that they state in a given language—natural and formal—how the system develops over time. In this respect, specifications denote high-level languages for solving problems without explicitly requiring the exact procedure to be followed. Computer programs, on the other hand, are *procedural*, as they determine in a step-wise manner how the physical computer is to behave.

To illustrate the difference between the specification and the computer program, take the description of the milk-delivery system again. This simulation specifies that a milk delivery truck must make one delivery at each store driving the shortest possible distance in total. According to Cantwell Smith, this is a description of what must happen, but not *how* it will happen. It is the computer program's responsibility to show how the delivery of milk actually takes place: "drive four blocks north, turn right, stop at Gregory's Grocery Store on the corner, drop off the milk, then drive 17 blocks north-east, [...]" (Cantwell 1985, 22).

Although correct in many respects, Cantwell-Smith's definition of 'computer program' is unpersuasive. The main concern here is that it fails to capture the difference between a step-wise procedure understood as syntactic formulae (i.e., the algorithm) from a step-wise procedure that puts the physical machine into the appropriate causal states (i.e., the computer process).[40] As we shall see shortly, this is not a harmless difference, but strikes at the core of many discussions on verification of computer software. In particular, failing to account for this distinction is at the basis of conflating the researcher's description of the behavior of the milk delivery system from the actual steps followed by the computer.

The notion of computer program,[41] then, needs to be further divided into two: the algorithm and the computer process.[42] Since we have discussed algorithms in some detail in the previous section, it is time to address the notion of computer process and its relation with algorithms.

Let me begin with the latter issue, as it helps to give meaning to the notion of computer process itself. As any researcher that has programed at least once in their life knows, in order to implement an algorithm on the computer, one needs to compile it first. Compiling basically consists of a mapping from an interpreted domain (i.e.,

---

[40]To complete the stages from specification to computer program, we should also add a host of intermediate steps, such as compiling the algorithm, allocation of memory, mass storage, etc. Because these intermediate steps do not constitute units of analysis for computer simulations, there is no need to discuss them in detail.

[41]Despite supporting this distinction, I keep the notion of a 'computer program' as a compact way to refer to the algorithm and the computer process altogether.

[42]The concept of "computer process" is, in turn, highly ambiguous since it may be used to refer to (i) encodings of algorithms, (ii) encodings of algorithms that can be compiled, (iii) encodings of algorithms that can be compiled and executed by a machine. James H. Moor has suggested up to five different interpretations (Moor 1988). See also (Fetzer 1988, 1058). My interpretation is similar to the third one.

the algorithm) into an interpreting domain (i.e., the computer process).[43] In other words, an algorithm is implemented as a physical process because the computer is able to interpret and execute the algorithm in the right way. Now, how is this possible?

At its core, every computer hardware consists of microelectronics built from billions of logical gates. These logic gates are the physical implementation of the logic operators 'and,' 'or,' and 'not' that, when combined, are enough to interpret all logic and arithmetic operations—and consequently, everything else.[44] At this level of description, all computers are basically the same, perhaps with the exception of the number of logical gates employed. However, this is how far the identity among computers goes, since at higher levels, not all computer share the same architecture.

A bottom-up schemata connects these logic gates with the computer process by involving a cascade of complex machine instructions and languages that 'talk' to each other. It starts with Microcode,[45] used as the hardware-level set of instructions able to implement higher-level machine code instructions, to the compiler, responsible for converting the instructions of the algorithm into a machine code, so that they can be read and executed as a computer process. Thus understood, a computer process runs on a physical computer because there are several layers of interpreters that translate a set of instructions into proper machine language.

Let us illustrate these points with a simplified case of the mathematical operation $2 + 2$ written in language C. Consider Algorithm 9.

Algorithm 9 A simple algorithm for the operation $2 + 2$ written in language C

```
void main()
{
return(2+2)
}
```

In binary code, the number 2 is represented by '00000010', whereas the plus operation is performed, for instance, by a Ripple-carry adder. The compiler then converts these instructions into machine code ready to be run on a physical computer. Once the computer process ends, the solution is displayed on the screen monitor, in this case 4, which is 00000100 in binary code.

Drawing from these considerations, we have now enough elements for presenting some key philosophical issues. Consider the following question: if computer processes are the physical realization of algorithms, which are abstract—and sometimes formal—how could the relation between one and the other be conceived? To

---

[43]On this point one could consult the work of Rapaport (1999; 2005).

[44]There are specific languages—so-called 'hardware description languages'—such as VHDL and Verilog which facilitate building these logic gates into the physical microcircuits (Cohn 1989; Ciletti 2010). These hardware description languages are, essentially, programming languages for the hardware architecture.

[45]Microcode was originally developed as a substitute to hardwiring the instructions for the CPU. In this way, the processor's behavior and programming is replaced by microprogrammed routines rather than by dedicated circuitry.

many, the sole purpose of algorithms is to prescribe the rules that the computer processes must follow on the physical computer. Let us note that understanding the relation between algorithms and computer processes in this way does not entail a return to Cantwell-Smith's notion of 'computer program,' for algorithms and computer processes are still two separate entities. There are several reasons that back up this claim. First, algorithms and computer processes are ontologically dissimilar. While algorithms are abstract entities, computer processes are causal in a straightforward physical sense. Moreover, while algorithms are cognitively accessible (i.e., researchers can understand and, to some extent, even follow the instructions set out in the algorithm), computer processes are cognitively opaque. A third reason is that a methodology of computer software requires acknowledging the existence of algorithms as intermediate between specifications and computer processes.[46] Thus understood, algorithms and computer processes are ontologically dissimilar, but they are epistemically equivalent. This means that the information coded in the algorithm is physically instantiated by the computer process, and considered to be epistemically on a par. The example is the addition of $2 + 2$ programed in Algorithm 9. Assuming a proper functionality of the compiler as well as the physical computer, the result of the computation of this algorithm is the actual addition, that is 4.

Accepting epistemic equivalence between algorithms and computer processes has some kinship with the verification debate in computer software. That is, given the formal verification of an algorithm, could researcher trust that the computer process also behaves in the way intended? Answering this question positively means that there is a—formal—method that ensures that computer processes behave as intended in the specifications and as programed in the algorithms. Answering negatively, on the other hand, raises the question of what grounds do researchers have to trust the results of computational processes. Let me now present the verification debate in more detail.[47]

Computer scientists and philosophers Hoare (1999) and Dijkstra (1974) believe that computer software is of a mathematical nature. In this context, computer programs can be formally verified, that is, the correctness of the algorithms can be proved—or disproved—with respect to certain formal properties in the specification, very much like a mathematical proof. The real difference with mathematics is that verification of computer software requires its own syntax. To this end, Hoare created his triples, which consist of a formal system—initial, intermediate, and final states—that strictly obey a set of logical rules. The Hoare triple has the form: $\{P\}$ C $\{Q\}$ where P and Q are assertions—*precondition* and *postcondition* respectively—and C is a command. When the precondition is met, the *command* establishes the postcondition. Assertions are formulae in predicate logic with a specific set of rules. One such rule is the empty statement axiom schema: $\overline{\{P\}Skip\{P\}}$; another is the assignment axiom schema: $\overline{\{P[E/x]\}x:=E\{P\}}$, and so forth (Hoare 1971).

---

[46]There have been some discussions on whether specifications—and algorithms—are executables, that is, whether a computer process can faithfully compute them. See Fuchs (1992).

[47]There is a large industrial and academic complex dedicated to verification and validation methods. Here I am interested in some original promoters and detractors. A complete discussion can be found in Colburn et al. (1993).

To Hoare and Dijkstra, computer programs are "reliable and obedient" (Dijkstra 1974, 608) as they behave exactly as instructed by the specification. The workload is, therefore, on the algorithm and how it correctly interprets the specification.[48] Once an algorithm is formally verified, the results of the computation will be as intended in the specification.

The opposite view draws from the fact that algorithms are ontologically different from computer processes, and concludes that they must also be epistemically different. While the former are indeed mathematical expressions suitable to mathematical—or logical—verification, the correctness of the latter can only be tackled by using empirical methods. This is the claim of the philosopher James Fetzer (1988), who believes that computer processes could hold an interpretation as causal in nature and therefore fall outside the scope of logical and mathematical methods. This means that, when the algorithm is implemented on the physical machine where causal factors are at play, the whole computer program becomes somehow 'causal.' His argument closes with the claim that formal verification is therefore impossible in computer science, and validation methods like testing must be granted a more important place.[49]

Many computer scientists and philosophers vehemently objected to Fetzer's argument, accusing him of not understanding the basics about the theory of computing.[50] Moreover, if Fetzer's argument is right, they argued, then the same should apply to mathematics since some kind of physical medium—our brain, a calculator, an abacus—is always needed for calculation and proof. Fetzer's argument, then, only holds if we acknowledge that there is a qualitative difference between the physical medium used for implementing an algorithm (i.e., the physical computer) and the physical medium used by a mathematician making a proof (i.e., our brain) (Blanco and García 2011).

Despite these objections, I believe Fetzer is right on two accounts. First, he is right in that algorithms and computer processes cannot be conceptualized as the same mathematical entity, but rather require different treatment. I have already mentioned this point earlier, when I ontologically separated algorithms as abstract and

---

[48]Dijkstra is much concerned with programmers and their education. At the time, the curriculum on formal verification was almost non-existent.

[49]I am of course simplifying the debate. Fetzer's argument is certainly more elaborate and worth studying in itself. To begin with, he makes a series of differences regarding the machine on which an algorithm and a process are executed (e.g., abstract machines, physical machines), differences that I have ignored here. Furthermore, when arguing for the possibilities of formal verification, Fetzer makes an essential distinction between *pure* mathematics and *applied* mathematics, the latter requiring the interpretation of a physical system (Fetzer 1988, 1059). As he puts it "if the function of a program is to satisfy the constraints imposed by an abstract machine for which there is an intended interpretation with respect to a physical system, then the behavior of that system cannot be subject to conclusive absolute verification but requires instead empirical inductive investigation to support inconclusive relative verifications." (Fetzer 1988, 1059–1060).

[50]A set of outrage letters were sent to the editor-in-chief of Communications of the ACM, Robert L. Aslzedzurst, immediately after Fetzer's article was published. Fortunately, the letters were published with an answer from the author, which not only shows the democratic spirit of the editor, but the fascinating reaction of the community towards this issue.

formal entities from computer processes as causally related. He is wrong, however, in thinking that these are grounds for rejecting formal verification and epistemic equivalence. Nowadays, many algorithms are formally verified (e.g., cryptographic protocols, and security protocols) and, when run on the computer, the computer processes are treated as epistemically equivalent to such algorithms.

Second, Fetzer highlights the role of validation—or testing—as being more relevant than originally thought to be. I fundamentally agree with Fetzer on this point. Validation methods cannot be absorbed and replaced by formal verification, even when the latter is possible. In fact, in the context of computer simulations, a combination of verification and validation methods are heavily used for entrenching the correctness of results. These issues are the object of Chap. 4. As we shall see there, verification methods focus on the relation model-specification, and thus it goes beyond 'formal' verification; validation methods, on the other hand, focus on the relation model-world, and thus are fundamental for ensuring that our model accurately represents the target system.

Finally, there is a further assumption that we need to mention. For our present purposes, I assume that there are no miscalculations or mathematical artifacts of any kind that the computer process introduces in the results. This presupposition is philosophically harmless and technically achievable. It follows, then, that the equations programed in the algorithm are reliably solved by the computer process, and that the results relate to the specification and the algorithm.

Figure 2.1 summarizes the three units of computer software and their connections. At the top level there is the specification, where decisions for the computer software are made and integrated altogether. The algorithm is the set of instructions that interprets the specification and prescribes how the machine should behave. As mentioned before, I call the pair <specification, algorithm> the *simulation model*. A simulation model, therefore, encompasses all the relevant information about the target system, and in this respect, is the most transparent unit of the computer simulation. Finally, there is the computer process as the semantic implementation of the algorithm on the physical computer.

Fig. 2.1 A general schema
of computer software.
Printed in (Durán 2014)

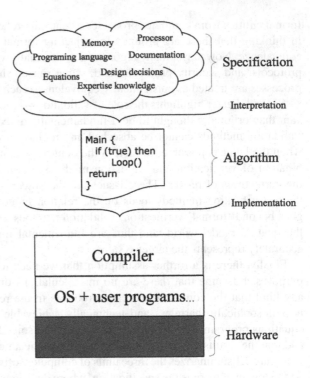

2.3 Concluding Remarks

This chapter focused on computer simulations as new kinds of units of analysis.
By identifying and reconstructing the specification, the algorithm, and the computer
process, we are in a position to better understand the nature of computer software in
general and computer simulations in particular.

The lesson to take home is that computer simulations are very complex units of
analysis, ones that bring together design and decision stages in the specification and
the algorithm, with the implementation on the physical computer. Each one of these
stages brings forward a series of concerns, whether they are related to social aspects
of computer simulations, their technical possibilities, or philosophical questions.
Although this chapter discussed many of these concerns, much more can and needs
to be said, especially with a focus on the nature of computer simulations.

The results obtained here will follow us throughout the rest of the book. For
instance, in Sect. 5.1.1, I argue for the possibilities for computer simulations to provide genuine understanding of the world by explaining it. As we shall see there,
the specifications and algorithms that constitute the simulation model enable the
explanatory force of computer simulations. Similarly, I will use these units of analysis explicitly during our discussion in Sect. 4.3.1, when I discuss errors in software
and hardware, and implicitly in Sect. 6.4 when I reference computer simulations as
a new paradigm of science.

Chapter 3
Units of Analysis II: Laboratory Experimentation and Computer Simulations

When philosophers fixed their attention on computer simulations, three different main lines of study emerged (Durán 2013a). The first line of study focuses on finding a suitable definition for computer simulations. A fundamental step towards understanding computer simulations is, precisely, to better grasp their nature by approaching a definition. This was the topic of our first chapter, where we tracked back definitions to the early 1960s.

The second line of study relates computer simulations with other units of analysis more familiar for researchers, such as scientific models and laboratory experimentation. Since this relation is established on a comparative basis, the kind of questions raised in this context are, among others, 'are computer simulations a form of scientific models, or are they forms of experimentation? What sort of knowledge should researcher expect by using a computer simulation, in comparison with the sort of knowledge obtained by using scientific models and experiments?' Comparing computer simulations with scientific models was the subject of Chap. 2, where I discussed their main constituents (i.e., specifications, algorithms, and computer processes). This chapter deals with comparing computer simulations with laboratory experimentation. In this respect, the main questions here are 'can computer simulations produce the sort of knowledge about the world that laboratory experimentation does? In what respects are computer simulations more—or less—fit to render reliable knowledge about the empirical world?' These questions have been at the heart of many philosophical debates about and around computer simulations. Here, I am interested in reconstructing some of this debate, its assumptions and implications.

Finally, the third line of study addresses computer simulations at face value, asking about their epistemological power to provide knowledge and understanding of a given target system. This third line is independent from the other two insofar as it is neither interested in defining computer simulations nor establishing comparisons with more familiar ways of inquiring into the world. The remaining chapters are dedicated to fleshing out many of the issues brought up by computer simulations.

© Springer Nature Switzerland AG 2018
J. M. Durán, *Computer Simulations in Science and Engineering*,
The Frontiers Collection, https://doi.org/10.1007/978-3-319-90882-3_3

3.1   Laboratory Experimentation and Computer Simulations

Computer simulations have been routinely compared with laboratory experimentation, as in many cases they are used in similar ways and for similar purposes.[1] Experimentation is typically conceived as a multidimensional activity stemming not only from the complexity surrounding the empirical phenomena under study, but also from the practice of experimentation in itself, which is intricate in design and complex in structure. That is why, when philosophers talk about *experiments*, they are referring to a host of interwoven topics, methodologies, and ways of practicing science. Experiments, for instance, are used for observing processes, detecting new entities, measuring variables, and even in some cases for 'testing' the validity of a theory. The questions that guide this section are, then, to what extent can we say that computer simulations are epistemically close—or even superior—to experimentation? and what set of characteristics do make them two distinctive—or similar—practices? Let us begin by gaining a better grasp of what experimentation is.

The idea of experimenting with Nature can be tracked back to the early times of civilization. Aristotle recorded his observation of the embryology of the chick in his *Historia Animalium* (Aristotle 1965), facilitating our early understanding of chicken and human development. In fact, Aristotle's studies correctly deduced the role of the placenta and the umbilical cord in humans. Although his methodology is flawed in several ways, it nonetheless resembles much of the modern scientific method: observation, measurement, and documenting every stage of growth—three aspects of scientific practice still in use until today.[2] Now, despite its undeniable centrality in our modern understanding of the empirical world, laboratory experimentation has not always received the appreciation it deserves.

It was not until the arrival of *logical empiricism* in the 1920s and 1930s that experiments began to receive some attention in the general philosophy of science. To the logical empiricist, however, experimentation represented not so much a philosophical problem in itself, as a subsidiary methodology for understanding theory. In fact, the most important use of experiments was for the *confirmation* and *refutation* of a theory, the most pervasive philosophical issue at the time.

A few decades later, logical empiricism began to experience a number of objections and attacks from different flanks. One particular objection played a fundamental role in their demise, which later became known as the *underdetermination of theory by evidence*. At its heart, this objection states that the evidence collected from experimentation might be insufficient for the confirmation or refutation of a given theory at

---

[1] Allow me to make one terminological clarification and one delimitation of topic. The clarification is that I use interchangeably and without further discussion the notion of *laboratory experimentation* and *experiment*. The subtleties of the distinction are of no interest for our purposes. As for the delimitation, I leave out of consideration *field experiments* as they typically require a different philosophical approach.

[2] The modern experimental method, however, must be ascribed to Galileo Galilei.

a given time. Logical empiricists, then, had no option but to embrace experimentation as a genuine part of scientific and philosophical inquiry.

Robert Ackerman and Deborah Mayo, two major names in the philosophy of experimentation, refer to the era where experimentation is at the center of philosophical inquiry as *new experimentalism*.[3] New experimentalism, as presented, supplements the traditional, theory-based view of logical empiricism with a more experimental-based view of scientific practice.

Although the advocates of the new experimentalism are interested in different kinds of problems emerging from experiments and their practices, they all share the assertion that scientific experimentation is at the heart of much of our understanding of the empirical world. The philosopher of experimentation Marcel Weber proposes five general trends that characterize new experimentalism. First, experimentation is *explorative*, that is, it aims at discovering new phenomena and empirical regularities. Second, new experimentalists reject the view that observation and experimentation is guided by theory. They maintain that in an important number of cases, theory-free experiments are possible and do occur in scientific practice. Third, new experimentalism has given new life to the distinction between observation and experimentation. Four, advocates of new experimentalism have challenged the positivist idea that theories somehow relate to nature on the basis of experimental results. And fifth, it has been stressed that more attention must be given to experimental practice in order to answer questions concerning scientific inference and theory testing (Weber 2005).

The shift from a traditional 'top-down' schemata (i.e., from theory to the empirical world) into a 'bottom-up' conceptualization is the distinctive mark of new experimentalism. Even notions like *natural phenomenon* went through some transformations. Under the new experimentalist point of view, a phenomenon can be from the directly observable cars crashing in front of our houses, to the invisible microbes, astronomical events, and quantum world.

Researchers give the label 'experimentation' to a wide range of activities. Aristotle's *observation* of the chick is perhaps the most straightforward use of the term. There is another broader use of the term that involves *intervention* or *manipulation* of nature. The idea is very simple and appealing: scientists manipulate an experimental set-up *as if* they were manipulating the empirical phenomenon itself. Whatever the epistemic gain from the former is, it can be extrapolated to the latter. Under this notion several activities can be identified. One such is *discovering new entities*, a highly valuable occupation in scientific research. For instance, Wilhelm Röntgen's discovery of the X-Rays is a good example of discovering a new type or radiation by manipulating nature.

---

[3]The work of Ackerman can be found in Ackermann (1989), and of Mayo in (Mayo 1994). Let the reader be advised that I am skipping several years of good philosophy of experimentation. Of particular interest is Norwood R. Hanson, a philosopher of science and fierce opponent of the logical empiricism, who made fundamental contributions to the transition from experiments as a subsidiary methodology of theory, to experiments as objects of studies in themselves. For references, see Hanson (1958).

Another example of manipulating nature are some types of *measuring quantities*. For instance, measuring the speed of light in the middle 1800s required a beam of light to reflect onto a mirror a few kilometers away. The experiment was set up in such a way that the beam would have to pass through the gaps between teeth of a rapidly rotating wheel. The speed of the wheel, then, was increased until the returning light passed through the next gap and could be seen. A very clever solution used by Hippolyte Fizeau to improve the accuracy of past measurements.

Understanding experiments and experimental practice this way raises questions about the relation between experiments and computer simulations. Are they epistemically on a par? or does the capacity to manipulate the real-world give experiments an epistemic advantage over computer simulations? Perhaps the most celebrated criteria for analyzing computer simulations and laboratory experiments is the so-called *materiality argument*. At its basis, the materiality argument says that, in genuine experiments the same material causes are at work in the experimental set-up as well as in the target system; in computer simulations, on the contrary, there is a formal correspondence between the simulation model and the target system.

Thus understood, the materiality argument offers different forms of ontological and epistemological commitments. One such a form takes experiments to be made of the same material causes as the target system, while computer simulations only share a formal correspondence with such target system. Under this interpretation, inferences about the target system are more justified in an experiment than in a computer simulation.[4] Alternatively, an argument can be advanced where experiments are similar to computer simulations, and therefore inferences by both are equally justified.

In the following, I partly reproduce an article of mine published in 2013 where I discuss in detail different ways to understand the materiality argument and its impact in the epistemological evaluation of computer simulations. This article provides, I hope, a similar level of technical and philosophical detail as the book. Let me finally say that much more philosophical work on the relation between computer simulations and experimentation has been published after this article. Examples are the excellent work of Parke (2014), Massimi and Bhimji (2015), and more recently Beisbart (2017).

3.2 The Materiality Argument

Much[5] of the current philosophical interest in computer simulations stems from their extended presence in scientific practice. This interest has centered on studies of the experimental character of computer simulations and, as such, on the differences— and similarities—between computer simulations and laboratory experiments. The

---

[4]The most comprehensible reconstruction of the materiality argument is given by Parker (2009).

[5]The following text has been partly published in Durán (2013b). Published with the permission of *Cambridge Scholars Publishing*.

philosophical effort, then, has been primarily focused on establishing the basis of this contrast; specifically by means of comparing the epistemic power of a computer simulation with that of a laboratory experiment. The basic intuition has been that if computer simulations resemble laboratory experiments in relevant epistemic respects, then they too can be sanctioned as a means of providing understanding of the world.

The standard literature on the topic distinguishes computer simulations from laboratory experiments on both ontological and representational grounds. The fact that a computer simulation is an abstract entity, and therefore bears only a formal relation to the system being investigated, contrasts with a laboratory experiment, which typically has a causal connection to the target system. These ontological and representational differences have suggested to some philosophers that establishing external validity is a much more difficult task for computer simulations than for laboratory experiments. For others, however, it has been a motivation to reconsider experimental practice, and see it as a broader activity that also includes simulations as a new scientific tool. These two approaches, I claim, share a common rationale that imposes restrictions on the epistemological analysis of computer simulations.

The most well-known criterion for distinguishing between computer simulations and laboratory experiments is given by the so-called *materiality argument*. Parker has provided a helpful account of this argument:

> In genuine experiments, the same material causes are at work in the experimental and target systems, while in simulations there is merely formal correspondence between the simulating and target systems [...] inferences about target systems are more justified when experimental and target systems are made of the same stuff than when they are made of different materials (as is the case in computer experiments). (Parker 2009, 484)

Two claims are being made here. The first is that computer simulations are abstract entities, whereas experiments share the same material *substratum* as the target system.[6] The second, which is essentially epistemic, is that inferences about empirical target systems are more justified by experiments than by computer simulations due to the material relations that the former bears with the world.

Current literature has combined these two claims into two different proposals: either one accepts both claims and encourages the view that being material better justifies inferences about the target system than being abstract and formal (Guala 2002; Morgan 2005); or one rejects both claims and encourages the view that computer simulations are genuine forms of experimentation and, as such, epistemically on a par with experimental practices (Morrison 2009; Winsberg 2009; Parker 2009). I claim that these two groups of philosophers, that superficially seem to disagree, actually share a common rationale in their argumentation. Concretely, they all argue for ontological commitments that ground their epistemic evaluations on computer simulations. I will refer to this rationale as the *materiality principle*.

---

[6]Some of the terminology in the literature remains unspecified, such as 'material' causes or 'stuff' (Guala 2002). I here take them to mean *physical causal relations*, as described, for instance, by Dowe (2000). In the same vein, when I refer to *causes, causality*, and similar terms, they should be interpreted in the way here specified.

In order to show that the materiality principle is at work in most of the philosophical literature on computer simulations, I discuss three distinctive viewpoints, namely:

(a) Computer simulations and experiments are ontologically similar (both share the same materiality with the target system); hence, they are epistemically on a par (Parker 2009);
(b) Computer simulations and experiments are ontologically dissimilar. Whereas the former is abstract in nature, the latter shares the same materiality with the phenomenon under study; hence, they are epistemically different (Guala 2002; Giere 2009; Morgan 2003, 2005);
(c) Computer simulations and experiments are ontologically similar (both are model-shaped); hence, they are epistemically on a par (Morrison 2009; Winsberg 2009).

With these three viewpoints in mind, the materiality principle can be reframed from another perspective: it is due to the philosophers' commitment to the abstractness—or materiality—of computer simulations that inferences about the target system are more—or less thereof—justified than laboratory experiments.

The principal aim here is to show that philosophers of computer simulations do adhere, in one way or another, to the materiality principle. I am also interested in outlining some of the consequences of adopting this rationale. In particular, I am convinced that grounding the philosophical analysis on the materiality principle, as most of current literature seems to do, places a conceptual corset on the study on the epistemological power of computer simulations. The philosophical study of computer simulations must not be restricted to, not limited by, a priori ontological commitments. By analyzing themes in the literature, then, I show that the materiality principle does not engender a helpful conceptualization of the epistemic power of computer simulations.

The following sections are divided in a way that corresponds to the three uses of the materiality argument listed above. The section entitled *the identity of the algorithm* discusses option (a); the section entitled *material stuff as criterion* addresses option (b), which comes in two versions, the *strong version* and the *weak version*; and finally option (c) is addressed in the section entitled *models as (total) mediators*.

3.2.1   The Identity of the Algorithm

Wendy Parker's formulation of the materiality argument has a prominent place in the recent literature on computer simulation. Following (Hartmann 1996), Parker defines a computer simulation as a time-ordered sequence of states that abstractly represents a set of desired properties of the target system. Experimentation, on the other hand, is the activity of putting the experimental setup into a particular state by means of intervening in it, and studying how certain properties of interest in the setup change as a consequence of that intervention (Parker 2009, 486).[7]

Parker's goal is to show that computer simulations and experiments share the same ontological basis, and to use this basis as justification for the claim that computer simulations and experiments are epistemically on a par. To her mind, the central problem is that current definitions of computer simulation do not qualify as an experiment because they lack the crucial intervening mechanisms. Indeed, it is the abstract character of the model that prevents computer simulations from serving as intervening systems. The solution to this issue consists in construing the notion of *computer simulation studies* as a computer simulation where an intervention is made into the physical computer itself. So defined, a computer simulation study does qualify as an experiment.

> A *computer simulation study* [...] consists of the broader activity that includes setting the state of the digital computer from which a simulation will evolve, triggering that evolution by starting the computer program that generates the simulation, and then collecting information regarding how various properties of the computer system, such as the values stored in various locations in its memory or the colors displayed on its monitor, evolve in light of the earlier intervention. (Parker 2009, 488)

The notion of *intervention* is now redefined as the activity of setting the initial state of the computing system and triggering its subsequent evolution. Thus understood, a computer simulation study is an experiment in a straightforward sense, for now the system intervened is the programmed digital computer (Parker 2009, 488). On this basis, Parker claims that there is ontological equivalency between computer simulations and experiments, and this in turn allows her to claim an equivalency in their epistemic power.

Notably, she does not explain what it means for a computer simulation study to be epistemically powerful. Instead, she limits the argument to asserting that an epistemology of computer simulations should reflect the fact that it is the observed behavior of the computer system that makes them experiments on a real material system—and therefore epistemically powerful.

The influence of the materiality principle can be now made more explicit. First, Parker seems to require that the digital computer is the '*substratum*' for the system being simulated, since this allows her to claim ontological equivalence between computer simulation studies and experiments. Furthermore, since the computer simulation study is the activity of putting the physical computer into an initial state, triggering the evolution of the simulation, and collecting physical data as indicated by prints-outs, screen displays, etc. (Parker 2009, 489), then the epistemic value of computer simulation studies also corresponds to that of experiments. In this sense, the evolution in the behavior of the programmed computer represents material features of the phenomenon being simulated. Finally, the researcher's understanding of such a phenomenon is justified by its evolution on the physical computer. Computer

---

[7]Intervention is conceived of as the manipulation of physical causal relations in the experimental setup.

simulation studies and experiments are, then, ontologically on a par, and so is their epistemological power.

Here I have briefly outlined Parker's main claims. The problem with her account, I believe, is that it is still not clear which are the reasons for considering the materiality of the digital computer as the relevant player in the epistemology of computer simulations. Let me put this concern in other terms. To my mind, Parker's motivations are to subvert the materiality argument by showing that computer simulations and experiments are ontologically on a par—and so is their epistemic power. This move, as I have argued, is grounded on a rationale behind the same materiality argument that she is trying to overthrow. The question, then, is what role does the materiality of the digital computer play in the evaluation of the epistemic power of computer simulation studies? Let me now offer three possible interpretations to this question.

First, Parker takes the materiality of the digital computer as playing the fundamental role of 'bringing about' the target system (i.e., brings into causal existence the phenomenon being simulated). In other words, the behavioral changes that the scientist observes in the physical computer are instantiations of the representations built into the computer simulation. Such representations are, naturally, representations *of* a target system. In this way, the physical computer behaves *as if* it were the empirical phenomenon being simulated in the programmed computer. In this respect, Parker says "[t]he experimental system in a computer experiment is the programmed digital computer (a physical system made of wire, plastic, etc.)" (Parker 2009, 488–489). It is not clear to me whether Parker is using a metaphor or, instead, she urges us to take this quote literally. In Duràn (2013b, 82), I refer to this interpretation as the 'phenomenon in the machine' and I show how it is technically impossible to obtain.

A second possible interpretation is that the system of interest is the physical computer itself, regardless of the represented empirical system. In this scenario, the researcher runs her simulations as usual, only paying attention to the changes in the behavior of the physical computer. These behavioral changes become the substance of the scientists inquiry, whereas the target system is only regarded as the initial point of reference for the construction of the simulation model. In this context, the researcher learns first and foremost from collecting information on the properties of the physical computer (i.e., the values in its memory and the colors on the monitor (Parker 2009, 488)). If this is the correct interpretation, then Parker must show that the scientist can cognitively access the various physical states of the computer, something that she fails to do. Philosophers have discussed whether it is possible to access different locations inside a computer (e.g., the memory, the processor, the computer bus, etc.) and the general agreement is that these locations are cognitively inaccessible for the unaided human. There is a guiding principle of epistemic opacity ascribed to computational processes which rules out any possibility of cognitively accessing the internal states of the physical computer (see my discussion in Sect. 4.3). Moreover, even if scientists could actually access these locations—say, if they were aided by another computer—it is still unclear why accessing these locations would be of any relevance for understanding the results of a computer simulation.

A third interpretation is that Parker takes the materiality of the physical computer to play some relevant role in the interpretations of results (Parker 2009, 490).

Under this interpretation, hardware failure, round-off errors, and analogous sources of miscalculation affect the results of the simulation in different ways. Since this is true of computers and of computation, then Parker's claim must be that the physical computer affects the final results of a computer simulation and, therefore, their epistemological assessment. If this is the correct interpretation, then I believe she is right. In Chap. 4, I present and discuss how researchers could know that the results of computer simulations are correct despite of the many sources of errors involved in the computation.

3.2.2   Material Stuff as Criterion

The advocates to the 'material stuff as criterion' are perhaps the best interpreters of the materiality argument. According to this view, there are fundamental and irreconcilable ontological differences between computer simulations and experiments, the latter being epistemically superior. There are two versions of this account: a *strong version* and a *weak version*.

The strong version holds that the causal relations responsible for bringing about the phenomenon must also be present in the experimental setup. This means that the experiment must replicate the causal relations present in the empirical system. According to the strong version, then, the experiment is a 'piece' of the world.

Take as an example a beam of light used for understanding the nature of the propagation of light. In such a case, the experimental setup is identical to the target system; that is, it simply is the empirical system under study. It follows that any manipulation of the experimental setup does address the same causes as the phenomenon, and that an insight into the nature of light can be delivered by our understanding of the controlled experiment (i.e., the beam of light (Guala 2002)).

Applied to computer simulations, the strong version takes it that the mere formal correspondence between the computer and the target system provides a sufficient basis for downplaying their status as epistemic devices. If there are no causal relations present, then the epistemic power of inferences thereby made about the world is downgraded.

The weak version, on the other hand, relaxes some of the conditions imposed by the strong version on experimentation. According to this view, a controlled experiment requires only the set of relevant causal relationships that bring the phenomenon about. In this vein, the proponents of the weak version do not commit themselves to a complete reproduction of the phenomenon under study, as the strong version does, but rather to the set of relevant causes that characterize the behavior of the phenomenon.

Let us illustrate the weak version with a simple example: a ripple-tank can be used as a material representation of light, thus providing insight into its nature as a wave. To the proponent of the weak version, it is enough to have a representative collection of causal correspondences between the experimental setup and the target system in order for the former to provide some insight into the latter. The relation between the experiment and the real-world phenomenon is, then, one of a subset of all causal relations. A cloud chamber detects alpha and beta particles, just as a Geiger counter can measure them, but neither instrument is a 'piece' of the phenomenon under study nor fully interacts with all kinds of particles. It follows that experimental practice, as exemplified by the detection and measurement of particles, depends on a complex yet partial set of all causal relations existing between the experimental setup and the target system.

Applied to the general evaluation of computer simulations, the weak version presents a more complex and rich picture, which affords of degrees of materiality being ascribed to computer simulations.

Despite these differences, however, both versions share the same viewpoint regarding computer simulations; namely, that they are epistemically inferior to experiments. This claim follows from the ontological conceptualization previously depicted, and stems from the same rationale that underlies the materiality principle.

3.2.2.1 The Strong Version

I belive that Francesco Guala champions the defense of the *strong version* when he assumes that an experiment reproduces the causal relations present in the phenomenon. In this respect, he assumes from the outset the existence of fundamental differences between computer simulations and experiments grounded on causality.

> The difference lies in the kind of relationship existing between, on the one hand, an experimental and its target system, and, on the other, a simulating and its target system. In the former case, the correspondence holds at a 'deep', 'material' level, whereas in the latter the similarity is admittedly only 'abstract' and 'formal' [...] In a genuine experiment the *same* 'material' causes as those in the target system are at work; in a simulation they are not, and the correspondence relation (of similarity or analogy) is purely formal in character. (Guala 2002, 66–67; emphasis mine).

To Guala's mind, changes in the materiality and their epistemological power can be understood in terms of sharing the 'same'—and 'different'—stuff. The case of the ripple-tank is paradigmatic in this respect. According to Guala, the media in which the waves travel are made of 'different' stuff: while one medium is water, the other is light. The ripple-tank, then, is a representation of the wave nature of light only because there are similarities in the behavior at a very abstract level (i.e., at the level of Maxwell's equations, D'Alambert's wave equation, and Hook's law). The two systems obey the 'same' laws and can be represented by the 'same' set of equations, despite their being made of 'different' stuff. However, water waves are not light waves, and a difference in the materiality presupposes a difference in the epistemic insight into nature (Guala 2002, 66).

The example of the ripple-tank is extrapolated onto the studies on computer simulations, for it allows Guala to claim that the ontological difference between experiments and simulations also grounds epistemological differences (Guala 2002, 63).

His loyalty to the materiality principle is thus unquestionable: there is a clear distinction between what we can learn and understand by direct experimentation, and what we can learn by a computer simulation. The epistemic payoff of the latter is less than the former and this is because, in this view, there is an ontological commitment to causality as epistemically superior that determines the epistemology of computer simulations.

Let me now consider a few objections to Guala's point of view. Parker has objected that his position is too restrictive for experiments, as well as for computer simulations (Parker 2009, 485). I agree with her on this point. Guala's conceptualization of experiments and computer simulations imposes artificial restrictions on both that are difficult to back up with examples in scientific practice. Moreover, and complementary to Parker's objection, I believe that Guala is adopting a perspective that takes both activities as chronologically mutually exclusive: that is, the computer simulation becomes a relevant tool when the experimentation cannot be implemented. STRATAGEM, a computer simulation of stratigraphy, provides us with an example here: when geologists are faced with difficulties in carrying out controlled experiments about strata formation, they appeal to computer simulations as the most efficacious replacement (2002, 68).[8] Such a tendency towards a disjunctive assessment of the two activities is a natural consequence of taking computer simulations to be epistemically inferior to experimentation. In other words, it is a natural consequence of adopting the materiality principle.

3.2.2.2 The Weak Version

For a proponent of the *weak version*, I turn to the work of Mary Morgan. She has presented the richest and most exhaustive analysis currently to be found in the literature regarding the differences between experiments and computer simulations.

Morgan is primarily concerned with so-called *vicarious experiments*, that is:

> Experiments that involve elements of nonmateriality either in their objects or in their interventions and that arise from combining the use of models and experiments, a combination that has created a number of interesting hybrid forms. (Morgan 2003, 271).

Having thus set out the features of vicarious experiments, she then turns to the question of how they provide an epistemic basis for empirical inference. Briefly, the more 'stuff' involved in the vicarious experiment, the more epistemically reliable it becomes. In simple words, degrees of materiality determine degrees of reliability. As Morgan comments: "on grounds of inference, experiment remains the preferable mode of enquiry because ontological equivalence provides epistemological power" (Morgan 2005, 326).

Morgan thus adheres to the weak version, because a vicarious experiment is characterized by different degrees of materiality, as opposed to the strong version

---

[8] Guala allows that experiments and computer simulations are appropriate research tools, *knowledgeproducers* as he calls them, although only for different contexts (2002, 70).

that holds that experiments must be a 'piece' of the world. In terms of the materiality principle, however, there are no fundamental differences between the two versions: she also considers ontology to determine the epistemological value of computer simulations. The difference lies, again, in the detailed analysis of the different kinds of experiments involved in scientific practice. Let me now briefly address her account.

As noted above, vicarious experiments can be classified according to their degree of materiality; that is, the different degrees to which the materiality of an object is present in the experimental setup. Table 3.1 summarizes four classes of experiments: *Ideal laboratory experiment* (also referred to as a *material experiment*), two kinds of *hybrid experiments*, and finally *mathematical model experiment*. As the table indicates, the classification is in terms of the kind of control exerted on the class of experiment, the methods for demonstrating the reliability of the results obtained, the degree of materiality, and the representativeness of each class.

The first and last classes are already well known to us: an example of an ideal laboratory experiment is the beam of light, for it requires effort by the scientist to isolate the system, rigorous attention to the control of the interfering circumstances, and intervention under these conditions of control. An example of the mathematical model experiment, on the other hand, would be the famous mathematical problem of the seven bridges of Königsberg; that is, a class of experiment whose control requirements are achieved by simplifying assumptions, whose demonstration method is via a deductive mathematical/logical method, and one whose materiality is, as expected, nonexistent (Morgan 2003, 218).

Among the number of ways in which these two classes of experiment differ, Morgan emphasizes those constraints imposed naturally via physical causality, and those imposed artificially via assumptions:

> The agency of nature creates boundaries and constraints for the experimenter. There are constraints in the mathematics of the model, too, of course, but the critical point is whether the assumptions that are made there happen to be the same as those of the situation being represented and there is nothing in the mathematics itself to ensure that they are (Morgan 2003, 220).

Hybrid experiments, meanwhile, can be conceived as experiments in-between the other two: they are neither fully material nor fully mathematical.[9] The class of *virtually experiments*, then, are understood as those "in which we have nonmaterial experiments on (or with) semimaterial objects," whereas *virtual experiments* are those "in which we have nonmaterial experiments but which may involve some kind of mimicking of material objects" (Morgan 2003, 216). Table 3.1 again summarizes the properties of all four kinds of vicarious experiments showing their representing and inference relations.

The differences between virtually and virtual experiments can be illustrated with the example of a cow hipbone used as surrogate for the internal structure of human

---

[9]Morgan says about this: "[b]y analyzing how these different kinds of hybrid experiments work, we can suggest a taxonomy of hybrid things in between that include virtual experiments (entirely nonmaterial in object of study and in intervention but which may involve the mimicking of observations) and virtually experiments (almost a material experiment by virtue of the virtually material object of input)" (Morgan 2003, 232).

Table 3.1 Types of experiment: Ideal laboratory, hybrids, and mathematical models with representing relations (Morgan 2003, 231)

	Ideal lab experiment	Hybrid experiments		Mathematical model experiment
		Virtually	*Virtual*	
Controls on:				
Inputs	experimental	experimental on	assumed	assumed
Intervention	experimental	inputs; assumed	assumed	assumed
Environment	experimental	on intervention and environment	assumed	assumed
Demonstration method	experimental in laboratory	simulation: experimental/ mathematical using model object		deductive in model
Degree of materiality of:				
Inputs	material	semimaterial	nonmaterial	mathematical
Intervention	material	nonmaterial	nonmaterial	mathematical
Outputs	material	nonmaterial	non- or pseudo-material	mathematical
Representing and Inference Relations	represent*ative of* ... ... to *same* in world represent*ative for* ... ... to *similar* in world...		representation of... ... back to *other* kinds of things in the world	

bones. For carrying out such an experiment, there are typically two alternatives: one can use a high-quality 3-D image of the hipbone that creates a detailed map of the bone structure, or, alternatively, a computerized 3-D image of the stylized bone; that is, a computerized 3-D grid representing the structure of the stylized bone. According to Morgan, the 3-D image has a higher degree of verisimilitude to the structure of the real hipbone because it is a more faithful representation of it, as opposed to the mathematization represented by the computerized 3-D grid (Morgan 2003, 230). The former is referred to as *virtually an experiment*, whereas the latter are called *virtual experiments*.

Now, what are the differences among all these different kinds of experiments? As shown in Table 3.1, whereas a virtually experiment is semi- or nonmaterial, an ideal laboratory experiment is strictly material. Also the demonstration methods are also significantly different. The distinction between a virtual experiment and a mathematical model, on the other hand, seems to be located solely in the method of demonstration, which is experimental for the former and deductive for the latter. Morgan also shows how models of stock market prices, despite being mathematical models simulated on a computer, can be categorized as a virtual experiment on account of the input data and the observation of results (Morgan 2003, 225). The boundaries between all four classes of experiment, however, are unfixed and dependent on factors external to the experiment in question. For instance, if a 3-D grid of the cow bone makes use of real measurements of the cow bone as input data, then what was originally a virtual experiment becomes virtually an experiment.

The epistemological analysis is a function of the degree of materiality of the class of experiment: "ontological equivalence provides epistemological power" (Morgan 2005, 326), as Morgan indicates. Back inference to the world from an experimental system can be better justified when the experiment and the target system are of the same material. As Morgan explains: "the ontology matters because it affects the power of inference" (Morgan 2005, 324). A computer simulation, for instance, cannot test theoretical assumptions of the represented system because it has been designed for delivering results consistent with built-in assumptions. A laboratory experiment, on the other hand, has been explicitly designed for letting the facts about the target system talk by themselves. According to Morgan, then, it is the material *substratum* underlying an experiment that is responsible for its epistemic power. Hence, the ideal laboratory experiment is epistemically more powerful than a virtually experiment; in turn, a virtually experiment is more powerful than a virtual experiment, and so on. Since computer simulations can only be conceived as hybrid experiments or as mathematical experiments, it follows that they are always less epistemically powerful than ideal laboratory experiments. To Morgan's mind, therefore, there are *degrees of materiality* that determine the *degrees of epistemic power*.

In this context, Morgan uses the terms *surprise* and *confound* to depict the epistemic states of the scientist regarding the results of a computer simulation and of a material experiment, respectively. Results of a computer simulation can only surprise the scientist because its behavior can be traced back to, and re-explained in terms of, the underlying model. A material experiment, on the other hand, can surprise as well as confound the scientist, for it can bring up new and unexpected patterns of behavior inexplicable from the point of view of current theory (Morgan 2005, 325), (Morgan 2003, 219). The materiality of the experiment, then, works as the epistemic guarantee that the results may be novel, as opposed to the simulation, which takes results as capable of being explained in terms of the underlying model. This shows how Morgan's ideas regarding experiments and computer simulations bear the stamp of the materiality principle. It exhibits the same rationale, putting materiality as the predominant feature for epistemic evaluation.

Despite Morgan's strong emphasis on the place that materiality has in the discovery of new phenomena, there are examples of virtual experiments whose epistemic power is clearly superior to any ideal laboratory experiment. Take as a simple example the dynamics of the micro fracture of materials. It is impossible to know anything about micro fractures without the aid of computers. Indeed, only the computational efficiency of finite element methods and multi-scale strong discontinuity can tell us something about the micro fractures of materials (Linder 2012). The lesson here is that understanding something about the world does not necessarily comes from material experiments, or from any degree of materiality whatsoever. Neither a field experiment nor a high-definition 3-D image would provide the understanding about the dynamics of micro fractures that can be provided by an accurate mathematical model.

3.2.3 Models as (Total) Mediators

The last account in my list is the one I called 'models as (total) mediators.' As the title suggests, this account is directly influenced by Morgan and Morrison's *Models as* Morrison (2009). The book is a defense of the mediating role of models in scientific practice, as it considers that scientific practice is neither driven by theories, nor is purely about direct manipulation of real-world phenomena. Instead, scientific practice needs the mediation of models in order to be successful in achieving its goals. A theory, then, cannot be directly applied to the phenomenon, but only by means of the mediation of a model; similarly, in experimental practice, models render data from measurements and observations in a form that is available for scientific use. In the following, I focus on the mediating role of models in experimental practice, since the proponent of the models as (total) mediators approach is more interested in analyzing computer simulations in the light of experiments. I will thus leave the mediating role of models in the context of theory unanalyzed (see my discussion in the previous chapter).

According to the proponent of the models as (total) mediators account, experimental practice consists in obtaining, by means of manipulating phenomena, data that informs us about certain properties of interest. This data, however, is in such a raw state that it is impossible to consider it reliable or representative of the properties measured or observed. Rather, for these raw data to be of any scientific use, it is necessary to further process it by filtering out noise, correcting values, implementing error-correcting techniques, and so forth. These correcting techniques are conducted by theoretical models and, as such, are responsible for rendering reliable data.

Scientific practice, then, is conceived as strongly mediated by models; and scientific knowledge is no longer obtained uniquely by our intervention into the world, but also by the conceptual mediation that the model/world relation represents. In this vein, the epistemic analysis is now concerned with the data filtered out, corrected, and refined by models, rather than the raw data collected by directly manipulating the real-world.

Computer simulations should easily fit into this new image of scientific practice. One might think that since they are conceived as models implemented on the digital computer, then their results must be data produced by a reliable model in a straightforward sense. Unfortunately, this is not what the proponent of models as (total) mediators has in mind. To her, it is correct to say that computer simulations are models running on a digital computer, and it is also correct to say that there is no intervention into the world in the empiricists sense. Nevertheless the data obtained by running a simulation are 'raw' in a similar sense as the data collected by a scientific instrument.[10] The reason for this is that there are material features of the target system that are being modeled into the simulation, and thus represented in the final simulated data (Morrison 2009, 53). Simulated data, then, need to be post-processed

---

[10]In order to keep these two notions of data separate, I will continue referring to data collected by the scientific instrument as 'raw data,' while I will refer to the data obtained by running the computer simulation as 'simulated data.'

by a further theoretical model, just in the same way as raw data. In other words, simulated data must also be filtered, corrected, and refined by another set of models in order to produce data that can be reliably used in scientific practice. Ontologically speaking, then, there are no differences between data produced by a scientific instrument and data produced by a computer simulation. These results give purchase to the claim that there are no epistemic differences between these two kinds of data either.

Let me now elaborate a bit more on these points. In 2009, Margaret Morrison published a fundamental contribution to the debate on measurement in the context of computer simulations. In that work, she claimed that certain types of computer simulations have the same epistemic status as experimental measurements precisely because both kinds of data are ontologically and epistemically comparable.

To illustrate this point, let us briefly consider her example of measuring the g force.[11] In an experimental measurement, Morrison argues, a scientific instrument measures a physical property up to a certain degree of precision, although such measurement will not necessarily reflect an accurate value of that property. The difference between *precision* and *accuracy* is of paramount importance for Morrison here: whereas the former is related to the experimental practice of intervening in nature—or computing the model in the simulation—the latter is related to the mediation of models as rendering reliable data. In this context, a precise measurement consists of a set of results wherein the degree of uncertainty in the estimated value is relatively small (Morrison 2009, 49); on the other hand, an accurate measurement consists of a set of results that are close to the true value of the measured physical property.[12]

The distinction between these two concepts constitutes the cornerstone of Morrison's strategy: data collected from experimental instruments only provide precise measurements of g, whereas reliable measurements must first and foremost be accurate representations of the value measured. It is in this context that Morrison considers that raw data must be post-processed in the search for accuracy (for the particular case of measuring g, Morrison proposes the ideal point pendulum as a theoretical model).

From Morrison's perspective, then, the reliability of the measured data is a function of the level of accuracy, which depends on a theoretical model rather than on the scientific instrument—or on the computer simulation.

> The level of sophistication of the experimental apparatus determines the precision of the measurement but it is the analysis of correction factors that determines the accuracy. In other words, the way modelling assumptions are applied determines how accurate the measurement of g really is. This distinction between precision and accuracy is very important—an accurate set of measurements gives an estimate close to the true value of the quantity being measured and a precise measurement is one where the uncertainty in the estimated value is small.

---

[11] Morrison also discusses the more sophisticated example of spin measurement (Morrison 2009, 51).

[12] The difference between precision and accuracy is framed by Franklin in the following example: a measurement of the speed of light, $c = (2.000000000 \pm 0.000000001) \times 10^{10}$ cm/s is precise but inaccurate, while a measurement $c = (3.0 \pm 0.1) \times 10^{10}$ cm/s is more accurate but has a lower precision (Franklin 1981, 367). For more details on accuracy and precision, see Chap. 4.

In order to make sure our measurement of g is accurate we need to rely extensively on information supplied by our modelling techniques/ assumptions (Morrison 2009, 49).

Computer simulations, just like scientific instruments, share the same fate of being precise but not accurate. For the latter, accuracy is tailored to the physical constraints in measuring the real-world; for the former, accuracy comes in the form of physical and logical constraints in the computation (e.g., round-off errors, truncation errors, and so forth). The precision/accuracy dichotomy, then, applies to computer simulations just as it does to experimental measurement, making both practices ontologically equal at the level of precise data, and epistemically equal at the level of accurate data. Thus understood, the materiality argument is also present here: equal ontology determines equal epistemology. And this was precisely the intention behind Morrison's analysis: "the connection between models and measurement is what provides the basis for treating certain types of simulation outputs as epistemically on a par with experimental measurements, or indeed as measurements themselves" (Morrison 2009, 36).

In the end, Morrison is applying a philosophy of modeling and experimentation onto a philosophy of computer simulations. This is also a consequence of following the materiality principle; that is, there is no analysis provided of computer simulations in itself, but only in the light of a more familiar philosophy. By making raw data and simulated data ontologically equal, and the post-processing a further epistemic step, Morrison is applying model techniques to computer simulations, regardless of the particularities of the latter. With this move in mind, Morrison also narrows down the class of computer simulations to those that are used as measuring devices; and in doing so, she is narrowing down the epistemic analysis to those simulations. The question that remains open is whether Morrison's strategy also works for all kinds of computer simulations (i.e., for those used with a purpose other than measuring).

3.3 Concluding Remarks

Many researchers make use of computer simulations as if they were reliable experimental devices. Such a practice presupposes that simulations are epistemically on a par with laboratory experimentation. In other words, computer simulations render at least as much and as qualitatively good knowledge about the surrounding empirical world as standard laboratory experimentation. But this is an unwarranted presupposition unless reasons are given that ground the epistemological power of computer simulations.

Following the discussions in this chapter, we noticed that the confidence of the researcher in using computer simulations might be affected by the 'materiality argument.' This argument says that our body of scientific knowledge is tailored to and depends on identifying the physical causal relations interacting in the empirical world. Computer simulations are, to many, abstract systems that only represent real-world phenomena. It then follows that laboratory experimentation, the traditional source that feeds into the body of scientific beliefs, still provides the most reliable path for knowing and understanding the world. The conclusion is then straightforward: researchers must prefer to run experiments than computer simulations, other things being equal.

But we must call into question whether this is really the case. There are many examples where computer simulations are in fact more reliable sources of knowledge about the world than traditional experimentation. Why is this the case? Why are researchers so confident about the use of computer simulations for providing insight into the world? These questions demand the kind of philosophical treatment on experimentation and computer simulations that this chapter provides.

The chapter presented three different views of how philosophers currently understand the epistemological study of computer simulations. I have shown that all three make use of the same rationale as the guide for their argumentation. I called this rationale the *materiality principle*, and I conceptualized it as the philosophers' commitment to an ontological account of computer simulations—and experimentation—that determines the evaluation of their epistemic power.

The general conclusion was that philosophers who accept the materiality principle are less likely to recognize what is distinctive about the epistemology of computer simulations than those who do not. The conclusion is modest, and aims at encouraging certain changes in the philosophical treatment of computer simulations. For instance, Barberousse et al. (2009) have made an important contribution to the notion of computer-simulated data, and Paul Humphreys has followed their work by analyzing the notion of data in more detail. Another excellent example is furnished by the role of computer simulation in climate modeling carried out by Parker (2014) and Lenhard and Winsberg (2010).

Despite these excellent works, much more needs to be done. To my mind, a potentially fruitful area of research is to reconsider certain classic topics in the philosophy of science through the lens of computer simulations. In this sense, a review of traditional notions of *explanation*, *prediction*, *exploration*, and the like might work as the starting point.[13]

Evidently, there is a way of doing philosophy of science that is strongly grounded on empirical inquiry exemplified by experimentation. The guiding epistemic principle is that the ultimate source of knowledge is given by an interaction with, and manipulation of the world. However, the continuous success of computer simulations is calling these principles into question: first, there is a growing tendency towards representing rather than intervening in the world; second, computational methods are pushing humans away from the center of the epistemological enterprise (Humphreys 2009, 616). The only definite conclusion is that the philosophical inquiry on the epistemological power of computer simulations has an arduous task ahead.

---

[13] I take on these enterprises in Chap. 5.

Chapter 4
Trusting Computer Simulations

Relying on computer simulations and trusting their results is key for the epistemic future of this new research methodology. The questions that interest us in this chapter are how do researchers typically build reliability on computer simulations? and what exactly would it mean to trust results of computer simulations? When we attempt to answer these questions, a dilemma is raised. On the one hand, it seems that a machine cannot be entirely reliable in the sense that they are not capable of rendering absolutely correct results. Several things could and typically do go wrong: from a systematic computational error to an unwary researcher that trips over a power cord. It is true that researchers develop methods and build infrastructure intended to increase the reliability of computer simulations and their results. However, absolute accuracy and precision are inherently a chimera in science and technology.

On the other hand, researchers trust the results of computer simulations—and it is important we keep reinforcing this trust—because they provide correct (or approximately correct) information about the world. With the aid of computer simulations, researchers are able to predict future states of a real-world system, explain why a given phenomenon occurs, and perform a host of standard as well as new scientific activities.

This dilemma brings forward a distinction that is not always made explicit in the scientific and technological parlance, but that is nonetheless central for assessing trust in computer simulations. The distinction is between *knowing* that the results are correct and *understanding* the results. In the former case, researchers know that the results are correct because the computer simulation is a reliable computational process.[1] But this does not mean that the researchers have also understood the results of the computer simulation. To understand them means to be able to relate the results with a larger corpus of scientific beliefs, such as scientific theories, laws and valid principles that govern nature. By knowing and understanding the results of computer

---

[1] Since I have also referred to computer simulations as methods, we could equally say that they are *reliable computational methods*. I use them indistinguishably.

© Springer Nature Switzerland AG 2018
J. M. Durán, *Computer Simulations in Science and Engineering*,
The Frontiers Collection, https://doi.org/10.1007/978-3-319-90882-3_4

simulations, researchers also know and understand something about the target system being simulated.

More often than not, researchers simply refer to the results of their simulations by saying 'we trust our results,' or 'we trust our computer simulations' meaning that they know the results are correct (or approximately correct) of a target system, that they understand why they are correct (or approximately correct) of a target system, or both. A chief aim in this chapter is to illuminate this distinction and its role in computer simulations.[2]

Philosophers have long argued that *knowledge* and *understanding* are two distinct epistemic concepts, and thus they must be treated separately. In the first sense above, researchers trust computer simulations when they know that the results are a good approximation to the real data measured and observed. In the second sense, researchers trust computer simulations when they understand the results and how they relate to the corpus of scientific beliefs. The difference can be illuminated with a simple example. One might know that $2 + 2 = 4$ without actually understanding arithmetic. An analogy with computer simulations can of course also be established. Researchers might know that the simulated trajectory of a given satellite under tidal stress is correct of the real trajectory without understanding why the spikes in the simulation occur (see Fig. 1.3).

In view of this distinction, two different questions are raised, namely, 'how do researchers know that the results of the simulation are correct of the target system', and 'what kind of understanding could be obtained?' To answer the first question, we need to discuss the available methods for increasing the reliability of computer simulations as well as the sources for errors and opacities that diminish such a reliability. To answer the second question, we need to address some of the many epistemic functions offered by computer simulations. The first question, then, is the subject of this chapter whereas the second question is the subject of the next chapter. Let us begin by making the distinction between knowledge and understanding more clear.[3]

---

[2]Let me make explicit that the following analysis has strong commitments to representation of a target system. The reason for taking this route is because most researchers are more interested in computer simulations that implement models that represent a target system. However, a non-representationalist viewpoint is also possible and desirable, that is, one that admits that claims such as 'the results suggest an increase of temperature in the Arctic as predicted by theory' and 'the results are consistent with experimental results,' are sound claims, instead of merely 'the results are correct of the target system.' This shift means that computer simulations are reliable processes despite of not representing a target system.

[3]Knowing and understanding are concepts that express our epistemological states and, in a sense, they can be taken to be 'mental.' If so, then neuroscience and psychology are disciplines better prepared to account for these concepts. Another way to analyze them consists in studying the concepts in themselves, showing their assumptions and consequences, and studying their logical structure. It is the latter sense in which philosophers typically discuss the concepts of knowledge and understanding.

4.1 Knowledge and Understanding

According to standard theories, knowledge consists in having reasons to believe a fact—also known as 'descriptive knowledge' or 'knowing that'. In more philosophical jargon, to know something is to have a true belief about that something, and to be justified in having such a belief.[4] Epistemologists, that is, philosophers specialized in theory of knowledge, lay out three general conditions for knowledge. Following the standard literature, the following schemata is in order: a subject S knows a proposition p if and only if:

(i) p is true,
(ii) S believes that p,
(iii) S is justified in believing that p.

The above schemata is known as 'Justified True Belief'—or JTB for short—, where the first premise stands for *true*, the second for *belief*, and the third for the *justification*. Epistemologists take it to be the minimal conditions for a subject to claim knowledge.

Let us now reconstruct JTB in the context of computer simulations. Let us call p the general proposition 'the results of a computer simulation are correct (or approximately correct) of the target system,' and S the researchers making use of computer simulations. It then follows that S has knowledge of p if:

(i) it is true that the results of a computer simulation are correct (or approximately correct) of the target system,
(ii) the researcher believes that it is true that the results are correct (or approximately correct) of the target system,
(iii) the researcher is justified in believing that it is true that the results are correct (or approximately correct) of the target system.

Condition (i), the truth condition, is largely uncontroversial. Most epistemologists agree that what is false cannot be known, and therefore there is little to debate around this condition. For instance, it is false to believe that Jorge L. Borges wrote *Principia Mathematica*, or that he was born in Germany. This is an example of the sort of thing that nobody would claim—or be in a position to claim—as knowledge. Similarly, no researcher would claim knowledge over results of computer simulations that depends upon basic arithmetic operations such as $a + b = (b + a) + 1$.

Condition (ii), the belief condition, is more controversial than the truth condition, but still largely accepted among epistemologists. It basically states that in order to know p, S is required to believe in p. Although a seemingly obvious claim, it has received several objections from philosophers that consider that knowledge without belief is also possible (Ichikawa and Steup 2012). Consider for instance a quiz where the student is asked to answer several question regarding Argentinean literature. One

---

[4]There are many good philosophical works on the notion of knowledge. The specialized literature includes Steup and Sosa (2005), Haddock et al. (2009) and Pritchard (2013).

such question is "where was Jorge L. Borges born?". The student does not trust her answer because she takes it to be a mere guess. Still, she manages to answer many of the questions well, including saying "Buenos Aires, Argentina". Does this student have knowledge about Argentinean literature? According to JTB, she does. This is an example brought up by Colin Radford in (1966), and counts as a fine piece of philosophical argumentation against JTB.

Now, neither the belief condition as presented by proponents of the JTB nor the criticism against it are issues that interest us here. This is so not only because of the inherent complexity of the subject matter that would take us too far from our main course, but mainly because there are good reasons for thinking that it is very unlikely that researchers get away with mere guesses about the results of computer simulations. First, it would be frankly quite amazing that somebody could guess the results of a computer simulation—in fact, in Sect. 4.3.2 I argue against this possibility. Second, there are dependable methods that reduce the possibilities and need of any epistemic luck about the correctness of the results. I then take it that the belief condition does not really concern us, and move to the real problem for computer simulation, that is, condition (iii), the justification condition.

The importance of condition (iii) is that a belief needs to be properly formed in order to be knowledge. A belief might be true and yet be a mere lucky guess, or even worse, induced. If I flip a coin and believe for no particular reason that it will land on tails, and if by mere chance the coin actually lands on tails, then there is no basis—other than chance—to say that my belief was true. Nobody can claim knowledge on the basis of mere chance. Consider now the case of a lawyer that employs sophistry to induce a jury into a given belief about a defendant. The jury might take that belief to be true, but if the belief is insufficiently well-grounded it does not constitute knowledge and thus lacks grounds for judging a person (Ichikawa and Steup 2012).

How could we accomplish justification in computer simulations? There are several theories of justification found in the specialized literature that come to our aid. In here, I am particularly interested in the so-called *reliabilism* theory of justification. Reliabilism, in its simplest form, takes that a belief is justified in the case that it is produced by a reliable process, that is, a process that tends to produce a high proportion of true beliefs relative to false ones. One way to interpret this in the context of computer simulations is to say that researchers are justified in believing that the results of their simulations are *correct* or *valid* with respect to a target system because there is a reliable process (i.e., the computer simulation) that, most of the time, produces accurate and precise results over inaccurate and imprecise ones.[5] The challenge now consists in showing how computer simulations qualify as a reliable process.

---

[5]Strictly speaking, p should read: 'the results of their simulations are correct', and therefore the researchers are justified in believing that p is true. To simplify matters, I will simply say that researchers are justified in believing that the results of their simulations are correct. This last sentence, of course, is taken to be true.

Alvin Goldman is the most prominent advocate of reliabilism. He explains it in the following way: "*reliability* consists in the tendency of a process to produce beliefs that are true rather than false" (Goldman 1979, 9–10; emphasis in orginal). His proposal highlights the place that a belief-forming process has in the steps towards knowledge. Consider, for instance, knowledge acquired by a reasoning process, such as doing basic arithmetic operations. Reasoning processes are, under normal circumstances and within a limited set of operations, highly reliable. There is nothing accidental about the truth of a belief that $2 + 2 = 4$, or that the tree in front of my window was there yesterday and, unless something extraordinary happens, it will be in the same place tomorrow.[6] Thus, according to the reliabilist, a belief produced by a reasoning process qualifies, most of the time, as an instance of knowledge.

The question now turns to what it means for a process to be reliable and, more specific to our interests, what this means for the analysis of computer simulations. Let us illustrate the first answer with an example from Goldman:

> If a good cup of espresso is produced by a reliable espresso machine, and this machine remains at ones disposal, then the probability that ones next cup of espresso will be good is greater than the probability that the next cup of espresso will be good given that the first good cup was just luckily produced by an unreliable machine. If a reliable coffee machine produces good espresso for you today and remains at your disposal, it can normally produce a good espresso for you tomorrow. The reliable production of one good cup of espresso may or may not stand in the singular-causation relation to any subsequent good cup of espresso. *But the reliable production of a good cup of espresso does raise or enhance the probability of a subsequent good cup of espresso. This probability enhancement is a valuable property to have.* (Goldman 1979, 28; emphasis mine)

The probability here is interpreted objectively, that is, as the tendency of a process to produce beliefs that are true rather than false. The core idea is that if a given process is reliable in one situation then it is very likely that, all things being equal, the same process will be reliable in a similar situation. Let it be noted that Goldman is very cautious in demanding infallibility or absolute certainty for the reliabilist account. Rather, a long-run frequency or propensity account of probability furnishes the idea of a reliable production of coffee that increases the probability of a subsequent good cup of espresso.

Borrowing from these ideas, we can now say that we are justified in believing that computer simulations are reliable processes if the following two conditions are met:

(a) The simulation model is a good representation of the empirical target system[7]; and

(b) The reckoning process does not introduce relevant distortions, miscalculation, or some kind of mathematical artifact.

At the very least both conditions must be met in order to have a reliable computer simulation, that is, a simulation whose results most of the time correctly represents

---

[6]Let us note that these examples show that a reliable process can be purely cognitive, as in a reasoning process; or external to our mind, as the example of a tree outside my window shows.

[7]As mentioned in the first footnote, we do not strictly need representation. Computer simulations could be reliable for cases when they do not represent, such as when the implemented model is well-grounded and it has been correctly implemented. I shall not discuss such cases.

empirical phenomena. Let me illustrate what would happen if one of the conditions above were not met. Suppose first that condition (a) is not met, as is the case of using the Ptolemaic model for representing the planetary movement. In such a case, although the simulation could render correct results, they do not represent any real planetary system and therefore the results could not be considered as knowledge of planetary motion. The case is similar if condition (b) is not met. This means that during the calculation stages there has been an artifact of some sort leading the simulation to render incorrect results. In such a case, the results of the simulation are expected to fail to represent the planetary motion. The reason is that miscalculations directly affect and downplay the degree of accuracy of the results.

In Sect. 2.2, I described with certain detail the three levels of computer software; namely, the *specification*, the *algorithm*, and the *computer process*. I also claimed that all three levels make use of techniques of construction, language, and formal methods that make the relations among them trustworthy: there are well-established techniques of construction based on common languages and formal methods that relate the specification with the algorithm, and allow the implementation of the latter on the digital computer. It is the totality of these relations that make the computer simulation a reliable process. In other words, these three levels of software are intimately related to the two conditions above: the design of the specification and the algorithm fulfill condition (a), whereas the running computer process fulfills condition (b). It follows that a computer simulation is a reliable process because its constituents (i.e., the specification, the algorithm, and the computer process) and the process of construing and running a simulation are based, individually and jointly, on trustworthy methods. Finally, from establishing the reliability of a computer simulation it follows that we are justified in believing (i.e., we *know*) that the results of the simulation correctly represent the target system.

We can now assimilate Goldman's realibilism into our question about knowledge in computer simulations: researchers are justified in believing that the results of a computer simulation are correct of a target system because there is a reliable process—the computer simulation—whose probability that the next set of results are correct is greater than the probability that the next set of results are correct given that the first results were just luckily produced by an unreliable process. In other words, results are to be trusted because computer simulations are reliable processes that produce, most of the time, correct (or approximately correct) results. The problem now lies in spelling out how to make computer simulations reliable processes. Let us now stop here and pick this issue back up in Sect. 4.2 where I discuss some of the conditions for reliability of computer simulations. Now it is time to discuss *understanding*.

At the beginning of this chapter, I mentioned that to know that $2 + 2$ is a reliable operation that leads to 4 does not entail an *understanding* of arithmetic. Understanding, unlike knowing, seems to involve something deeper and perhaps even more valuable that is comprehending that something is the case.

Why is an analysis on understanding important? The short answer is that scientific understanding is essentially an epistemic notion that involves scientific activities such as explaining, predicting, and visualizing our surrounding world. There is, however, general agreement that the notion of understanding is hard to define. We say that we 'understand' why the Earth revolves around the Sun, or that the velocity of a car could be measured by deriving the position of the body with respect to time. But finding the conditions under which we understand something are surprisingly more difficult than for knowing.

A first characterization takes understanding as the process of populating a coherent corpus of scientific true beliefs (or close to the true beliefs) about the real-world. Such beliefs are true (or close to the truth) in the sense that our models, theories, and statements about the world provide reasons to believe that the actual world is not likely to be significantly different (Kitcher 1989, 453).

Naturally, not all scientific beliefs are strictly true. Sometimes we do not even have a perfect understanding of how our scientific theories and models work, let alone a complete grasp of why the world is the way it is. For these reasons the notion of *understanding* must also allow some falsehoods. The philosopher Catherine Elgin has coined an adequate term for these cases; she calls them 'felicitous falsehoods' as a way to exhibit the positive side of a theory of not being strictly true. Such felicitous falsehoods are the idealizations and abstractions that theories and models purport. For instance, scientists are very well aware that no actual gas behaves in the way that the kinetic theory of gases describes them. However, the ideal gas law accounts for the behavior of gases by predicting their movement, and explaining properties and relations. There is no such a gas, but scientists purport to understand the behavior of actual gases by reference to the ideal gas law (i.e., to reference a coherent corpus of scientific beliefs) (Elgin 2007, 39).

Now, although any scientific corpus of beliefs is riddled with felicitous falsehoods, this does not mean that the totality of our corpus of beliefs is false. A coherent body of predominantly false and unfounded beliefs, such as alchemy or creationism, still does not constitute understanding of chemistry or the origins of beings, and it certainly does not constitute a coherent corpus of scientific beliefs. In this vein, the first demand for having understanding of the world is that our corpus is mostly populated with true (or close to the truth) beliefs.

Taken in this way, it is paramount to account for the mechanisms by which new beliefs are incorporated into the general corpus of true beliefs, that is, how it is populated. Gerhard Schurz and Karel Lambert assert that "to understand a phenomenon P is to know how P fits into one's background knowledge" (Schurz and Lambert 1994, 66). Elgin echoes these ideas when she says that "understanding is primarily a cognitive relation to a fairly comprehensive, coherent body of information" (Elgin 2007, 35).

There are several operations that allow scientists to populate our scientific corpus of beliefs. For instance, a mathematical or logical derivation from a set of axioms incorporates new well-founded beliefs into the corpus of arithmetics or logic, making them more coherent and integrated. There is also a pragmatic dimension that considers that we incorporate new beliefs when we are capable of using our scientific corpus of belief for some specific epistemic activity, such as reasoning, working with hypotheses, and the like. Elgin, for instance, calls attention to the fact that understanding geometry entails that one must be able to reason geometrically about new problems, to apply geometrical insight in different areas, to assess the limits of geometrical reasoning for the task at hand, and so forth (Elgin 2009, 324).

Here, I am interested in outlining four particular ways of incorporating new beliefs into the corpus of scientific knowledge. These are, by means of explanation, by means of prediction, by means of exploration of a model, and by means of visualization. To this end, I show how each of these epistemic functions work as a coherence making process capable of incorporating new beliefs into our scientific corpus of beliefs. In some cases, the process of population is rather straightforward. Philosophers working on scientific explanation, for instance, have largely admitted that the aim of explanation is precisely to provide understanding of what is being explained. The philosopher Jaegwon Kim says that "the idea of explaining something is inseparable from the idea of making it more intelligible; to seek an explanation of something is to seek to understand it, to render it intelligible" (Kim 1994, 54). Stephen Grimm, another philosopher of explanation, makes the same point with fewer words: "understanding is the goal of explanation" (Grimm 2010, 337). Explanation, then, is an important driving force for scientific understanding. We can understand more about the world because we can explain why it works the way it does, and thus populating our scientific corpus of beliefs. A successful account of explanation for computer simulations, then, must show how to render understanding by simulating a piece of the world. A similar argument is used for the other epistemic functions of computer simulations. This is, however, the subject for the next chapter.

4.2 Building Trust

Above, I stated that the challenge for claiming that the results of a computer simulation are trustworthy consists in showing that they are produced by a reliable process—namely, the computer simulation. The question that concern us now, then, is by what means could researchers entrench the reliability of computer simulations? Over the years, researchers have developed different methods that facilitate such purpose. In the following sections, I dedicate some time to discuss these methods and how they affect the researcher's trust in the results of computer simulations.

4.2.1 Accuracy, Precision, and Calibration

Let us first clarify three terms that are central for assessing the reliability of computer simulations. These are *accuracy*, *precision*, and *calibration*. Contemporary studies have generated two main bodies of research. On the one hand, there are mathematical theories of measurement whose main concern is the mathematical representation of quantities, standardization, units and systems, and methods for determining ratios and quantities. On the other hand, there are philosophical theories concerned with the methodological, epistemological, and metaphysical assumptions of measurement. Here we are of course interested in the latter.

In the sciences, researchers carry out measurements of a number of quantities of interest which can be more or less accurate of that quantity. What does this mean? Traditional accounts of measurement theory take that *accuracy* refers to the set of measurements that provide an estimated value close to the *true* value of the quantity being measured. In this respect, accuracy refers to whether the quantity has been correctly determined by comparison. For instance, the measurement of the speed of light $c = (3.0 \pm 0.1) \times 10^8$ m/s is accurate insofar as it is close to the true value of the speed of light in a vacuum, namely, 299,792,458 m/s.

Such conceptualization seems about right, except that it presupposes knowledge of the true value of the speed of light. That is to say, that researchers have access to the actual, fixed value of a quantity in nature, and that the means by which we access such a value is granted (i.e., unbiased). Unfortunately, this is not a realistic picture of standard measurement in science and engineering. Rather, it is more usual to see researchers struggling to secure a value by means of measuring techniques and instruments that inevitably will introduce some bias. For these reasons, researchers take it that, at best, they can only measure approximate values of a quantity. For example, in measuring the speed of light, researchers are idealizing the medium in which light travels, in this case, a vacuum. Other idealizations must also be in place, such as temperature and pressure, and the stability of units of measurement. Removing these idealizations would imply a more complex scenario where, most likely, researchers will never get to know the true value of the speed of light (Teller 2013).

It is due to philosophical arguments like this that our world becomes less certain. Of course, the fact that in many contexts researchers cannot measure the true value of a quantity does not mean that, for all scientific and engineering disciplines, measuring the true value of a quantity is impossible. In trigonometry, for instance, one can theoretically know the true value of each angle of a rectangle. However, for many scientific and engineering disciplines, it is really difficult—if not impossible—to talk of measuring the *true* value of (empirical) quantities. Even for cases where measurement of a value is obtained by theoretical means, there are reasons to believe that there are a set of assumptions that impose constraints on the measurement. Take the example of measuring a second. "Since 1967," says Eran Tal, "the second has been defined as the duration of exactly 9,192,631,770 periods of the radiation corresponding to a hyperfine transition of cesium-133 in the ground state (BIPM 2006). This definition pertains to an unperturbed cesium atom at a temperature of absolute zero. Being an idealized description of a kind of atomic system, no actual cesium atom ever satisfies this definition" (Tal 2011, 1086). The lesson to take home is that measuring a *true* quantity—especially in nature—is not possible, and this affects the definition of accuracy by comparison.

Another way to analyze accuracy consists in having a closer look at scientific and engineering practice, particularly at the design and use that researchers have for instruments. For instance, if the design and use of a thermometer warrants the attribution of temperature to an object within a narrow range, then we can say that such thermometer is accurate. Thus understood, an accurate measurement presupposes a complete access to the design of the instrument, the laws that govern it—in this case, thermodynamics—external conditions that might affect the instrument (e.g., a hand that warms up the thermometer), and the uses of the instrument.

A third way consists in taking that a measurement is accurate with respect to a given standard of measurement. The case of the measurement of the speed of light is one case in point. If it is established as standard that the speed of light is 299,792,458 m/s, then a measurement of $c = (3.0 \pm 0.1) \times 10^8$ m/s is accurate insofar as it is close to the standard value. The great advantage of making accuracy depend on measurement standards is that researchers do not need to presuppose a fixed value in nature, but rather a fixed value in time. This means, in turn, that different periods in the history of science and technology will give birth to new—and modify existing—measurement standards. A convincing example is furnished, again, by the speed of light. In 1907, the value of c was about $299,710 \pm 22$ km/s, but by 1950 the established result was $299,792.5 \pm 3.0$ km/s. Each one of these measurements, and the many that came after, depend on a different method and instrument that are typically tailored to a specific point in time. For this viewpoint, it is absolutely essential to combine the methods, the instruments, and the socio-technological time-frame in the conceptualization of the notion of accuracy.

Yet a fourth alternative way to understand accuracy is on multiple comparative bases with other scientific and engineering instruments. Thus, if one instrument's outcome agrees on the same value with other instruments, preferably of a different kind (e.g., mercury thermometer, infrared thermometer, digital thermometer), then it is said to be accurate (Tal 2011, 1087).

What count as accurate results, then, depends on measuring methods, standards, instruments, and of course, the scientific and engineering communities. In this respect, researchers can take results of a computer simulation as accurate if they are in close agreement to a value of a quantity obtained by reference to other values, provided a reasonably narrow range; by reference to a measurement standard, such as the speed of light or absolute zero; or by comparing the results with other scientific instruments—including, in many cases, other computer simulations. A good example of this last case is provided by Marco Ajelli and team (Ajelli et al. 2010), who compare side-by-side the results of two different simulations: a stochastic agent-based model and a structured metapopulation stochastic model. According to the authors, "the results obtained show that both models provide epidemic patterns that are in very good agreement at the granularity levels accessible by both approaches" (Ajelli et al. 2010, 1).

Accuracy is often related to *precision*, although a careful analysis shows that they must be kept differentiated. The analysis of correction factors and model assumptions determines the accuracy of a measurement, whereas it is the level of sophistication of the scientific instrument that determines the precision of a measurement. The classic example that helps set these two apart is that of target archery. The accuracy of the archer is determined by how closely the arrows cluster around the target's bull's eye. The precision of the archer, on the other hand, corresponds to how closely (or widely) the arrows are scattered. The tighter the arrows are, the more precise the archer. Now, if the distance from the bull's eye is considered too far, then the archer is to said to have little accuracy.

The distinction between precision and accuracy is very important. An accurate measurement gives an estimate close to the true value of the quantity being measured. In order to make a measurement accurate, researchers need to rely extensively on the information about their models and mathematical machinery. A precise measurement, on the other hand, is one where the uncertainty in the estimated value is reasonably small. In order to make a measurement precise, researchers rely on the sophistication of their instruments. As I show later, in computer simulations precision also depends on information about the researcher's model and mathematical machinery.

In measurement theory, *precision* refers to the set of repeated measurements under unchanged conditions for which the uncertainty of the estimated results is relatively small. In order to increase precision, there are a number of statistical methods that help in providing more precise results, such as the standard error and the standard deviation. A good example is again furnished by the measurement of the speed of light; a set of measurements of the speed of light c whose standard deviation is about $\pm 0.0000807347498766737$ is less precise than one whose standard deviation is about $\pm 0.0000008073474987667$, as the uncertainty of the estimated results is smaller. Thus understood, the sophistication of the scientific instrument (e.g., a laser for measuring the speed of light) determines the precision of the measurement of c, whereas the accuracy of the measurement depends on the analysis of correction factors, model assumptions, and the like.

In computer simulation, precision is an important element to care about. In typical circumstances researchers assume, rightly to my mind, that the computer as a physical machine is a reasonably precise instrument. That is to say that the results obtained from computing are distributed within an acceptable range (e.g., within a normal distribution). Since studies on precision also focus on sources of instrument imprecision, such as uncontrolled variations to the equipment, malfunctioning, and general failure, it is important to point out the same sources that affect computer simulations. The homology with computer simulations, then, can be established by factors such as controlling the overheating of computers, I/O device errors, and memory exceptions, among others.

Besides the hardware layer, computer simulations involve a software layer which can also introduce imprecisions into the computation. Usually, the limited amount of bits used to store a number lead to known—and unknown—truncation and rounding off errors. A simple example consists in storing "sin(0.1)" in IEEE single precision floating point standard.[8] If subsequent computations make use of this number, the error tends to be magnified risking precision in the computation. In order to make the measurement of sin(0.1) more precise, researcher usually improve their hardware basis—for instance, from a 64 bits to a 128 bits architecture. Of course, this only

---

[8]The example also works for showing the inaccuracies introduced by computing "sin(0.1) in IEEE single precision floating point.

means that the rounding off tends to be smaller—or have less of an impact in the final results—but are not necessarily eliminated. This is a reason why, in High-Performance Computing, hardware plays such a relevant role. Unfortunately, changes in the hardware can be quite expensive, so precision must be also dealt with by means of numerical analysis and good programing.

Finally, calibration—also known as 'tuning'—refers to the host of methods that allow small adjustments to be made to the model's parameters such that the degree of accuracy and precision desired reach a specific standard of fidelity. Calibration typically—although not always—occurs when a model includes parameters about which there is much uncertainty, and thus the value of the parameter is determined by finding the best way to fit the results with the available data. The parameter in question is then considered a free parameter that can be 'tuned' as needed. Calibration is then the activity of finding and adjusting values for the free parameter that are best for accounting for the available data.

In computer simulation, Marc Kennedy and Anthony O'Hagan identify two forms of calibration. First, 'calibration inputs' which are inputs that take fixed but unknown values for all the measurements and observations used for the calibration. Thus understood, calibration inputs are those inputs that we wish to learn about through the process of calibrating the simulation. Second, 'variable inputs' which consist of all other inputs whose value might change during the execution of a computer simulation. These generally describe the geometry, and initial and boundary conditions associated with specific aspects of the target system.[9] For both forms of calibration, there is a host of methods available including parameter estimations by nonlinear regression, Bayesian methods, and sensitivity analysis for evaluating the information content of data and identifying existing measurements that dominate model development.

Although part of the standard practice of computer simulations, calibration comes with significant drawbacks. A central concern with calibration is that it might result in data 'double-counted.' This is, data used for the calibration of the computer simulation can also be used for the evaluation of the accuracy of the results of a computer simulation. This concern, particularly present within the community of climate simulation, raises questions about circularity and self-confirmatory stances:

> some commentators feel that there is an unscientific circularity in some of the arguments provided by GCMers [general circulation modelers]; for example, the claim that GCMs may produce a good simulation sits uneasily with the fact that important aspects of the simulation rely upon [...] tuning. (Shackley et al. 1998, 170)

Another problem with calibration is the 'residual uncertainties' that come with it. As we have discussed, calibration consists of searching for a set of values of the unknown inputs such that the available data fits as closely as possible to the outputs of the model. These values serve several purposes, and they are all estimates to the true values of these parameters. The 'estimative' nature of these inputs entails a

---

[9]For more on this issue, see McFarland and Mahadevan (2008), Kennedy and O'Hagan (2001), and Trucano et al. (2006).

residual uncertainty on these inputs. In other words, calibration does not eliminate uncertainty, it simply reduces it. This fact must be recognized in subsequent analysis of the model.

The above efforts to characterize accuracy, precision, and calibration stem from considerations on the reliability of computer simulations and the subsequent trust that researchers have in their results. If computer simulations are reliable processes that render correct results, then we need a way to characterize what a 'correct' result would be. Thus, the reason for our recent discussion. As is standard in this book, we present and discuss, however briefly, philosophical problems attached to concepts and ideas. The next step is to give reasons for saying that computer simulations are reliable processes. To this end, I offer a brief discussion on the literature on verification and validation as the two most important methods for granting reliability.

4.2.2 Verification and Validation

Verification and *validation*[10] are the general names given to a host of methods used for increasing reliability in scientific models and computer software. Understanding their role, then, turns out to be essential for the assessment, credibility and power to elicit the results of computers. Now, when the interest is on computer simulations, verification and validation methods are tailored to specific tasks. In the following, I'll first discuss the generalities of verification and validation methods and discuss later in what respect they increase the reliability of computer simulations.

In order to account for the reliability of the computer simulations as well as to sanction the correctness of the results, researchers have at their disposal formal procedures (e.g., for confirming the correct implementation of the specification in the computer) and benchmarks (i.e., accurate reference values) against which to compare the results of the computation (e.g., against other sources of data). In *verification*, formal methods are at the center for the reliability of computer software, whereas in *validation*, benchmarking is responsible for confirmation of the outcomes (Oberkampf and Roy 2010, Preface). In other words, in verification methods, the relationship of interest is between the specification—including the model—and the computer software, whereas in validation methods the relationship of interest is between the computation and the empirical world (e.g., experimental data obtained by measuring and observing methods) (Oberkampf et al. 2003).[11]

Here are two definitions that are largely accepted and used by the community of researchers:

---

[10] Also known as 'internal validity' and 'external validity,' respectively.

[11] To be more fair with Oberkampf's, Truncano's, Roy's, and Hirsch's general proposal, I must also mention the analysis on uncertainty and how it propagates throughout the process of designing, programming, and running computer simulations. For more philosophical treatment on verification and validation, as well as concrete examples see Oreskes et al. (1994); Küppers and Lenhard (2005); Hasse and Lenhard (2017).

Verification: the process of determining that a computational model accurately represents the underlying mathematical model and its solution.

Validation: the process of determining the degree to which a model is an accurate representation of the real world from the perspective of the intended uses of the model. (Oberkampf et al. 2003)

Now, this way of depicting verification and validation is widely accepted and used, specially in many philosophical treatments of computer simulations. Eric Winsberg, for instance, takes it that *"verification*, [...] is the process of determining whether or not the output of the simulation approximates the true solutions to the differential equations of the original model. *Validation*, on the other hand, is the process of determining whether or not the chosen model is a good representation of the real-world system for the purpose of the simulation" (Winsberg 2010, 19–20). Another example of a philosopher discussing verification and validation for computer simulations is Margaret Morrison. Although she takes a wider definition of verification and validation methods, and even thinks that these two methods are not always clearly divisible, she nevertheless downplays the need for verification methods claiming that validation is more crucial of a method for assessing the reliability of computer simulation (Morrison 2009, 43).

The scientific and engineering communities, on the other hand, have a wider and more diverse set of definitions to offer, all tailored to the specificities of the systems under study.[12] Let us now discuss them separately and point out what is so specific about computer simulations.

4.2.2.1 Verification

Verification in computer simulations consists in making sure that the specification for a given simulation is correctly implemented as a simulation model. The literature provides several verification methods suitable for computer software in general, but there are two methods specifically important for computer simulations, namely, *code verification* and *calculation verification*.[13] Their importance lies in that both methods focus on the correctness of the discretization, a key element for implementing mathematical models as computer simulations.

Code verification is defined as the process of determining that the numerical algorithms are correctly implemented in the computer code, as well as of identifying potential errors in the software (Oberkampf et al. 2003, 32). In this respect, code verification provides a framework for developing and maintaining reliable computer simulation code.

William Oberkampf and Timothy Trucano have argued that it is useful to further segregate code verification into two activities, namely, *numerical algorithm*

---

[12]See Oberkampf and Roy (2010), 21–29 for an analysis on the diversity of concepts. Also, see Salari and Kambiz (2003), Sargent (2007), Naylor (1967a, b).

[13]Also referred to as *solution verification* in Oberkampf and Roy (2010), 26, and as *numerical error estimation* in Oberkampf et al. (2003, 26).

verification and *software quality engineering*. The purpose of numerical algorithm verification is to address the mathematical correctness of the implementation of all the numerical algorithms that affect the numerical accuracy of the results of the simulation. The goal of this verification method is to demonstrate that the numerical algorithms implemented as part of the simulation model are correctly implemented and functioning as intended (Oberkampf and Trucano 2002, 720).

Software quality engineering, instead, puts the emphasis on determining whether the simulation model produces the correct results—or approximately correct. The purpose of software quality engineering is to verify the simulation model and the results of the simulation on a specific computer hardware, in a specified software environment—including compilers, libraries, I/O, etc. These verification procedures are primarily in use during the development, testing, and maintenance of the simulation model (Oberkampf and Trucano 2002, 721).

On the other hand, *calculation verification* is defined as the method that prevents three kinds of errors: human error in the preparation of the code, human error in the analysis of the results, and numerical errors resulting from computing the discretized solution of the simulation model. A definition for calculation verification is the following:

> *Calculation verification*: the process of determining the correctness of the input data, the numerical accuracy of the solution obtained, and the correctness of the output data for a particular simulation (Oberkampf et al. 2003, 34)

Thus understood, calculation verification is the empirical side of verification. It is based on the comparison between the results of the simulation against highly accurate solutions of the scientific model. In a sense, calculation verification is similar to validation assessment insofar as both compare estimated results with correct results. It most commonly controls spatial and temporal convergence rates, iterative convergence, independence of solutions to coordinate transformations, and similar other processes (Oberkampf et al. 2003, 26).

4.2.2.2 Validation

The process of *validation* (also known as *testing*) consist in showing that the results of the simulation correspond, more or less accurately and precisely, to those obtained by measurement and observation of the target system. Oberkampf and Trucano highlight three key aspects of validation:

> (i) quantification of the accuracy of the computational model by comparing its responses with experimentally measured responses,
>
> (ii) interpolation or extrapolation of the computational model to conditions corresponding to the intended use of the model, and
>
> (iii) determination if the estimated accuracy of the computational model, for the conditions of the intended use, satisfies the accuracy requirements specified. (Oberkampf and Trucano 2008, 724)

Although validation methods come naturally to many experimenters, as they expect to reproduce—as opposed to represent or imitate—a piece of the world,[14] they are rather a complex matter in the context of computer simulations. Here is when some concerns appear regarding the dependability of validation.

A chief problem stems from the fact that most validation methods are inductive, and thus must be expected to face typical problems of induction. The issue here is that the method only allows validation of a model up to a certain point, and therefore complete validation is absolutely impossible due to the large number of comparisons needed—not to mention the improbability of having all the possible results at hand. This is a reason why validation is known, mostly among computer scientists, as a method for *detecting* the presence of errors, but not designed for *establishing* their absence.[15]

Another problem is that validation depends on the capacity of comparing computer simulation results with *empirical* data. Such a dual relation obviously requires the presence of both, the results of the simulation and data gathered from an empirical source. This leads to the exclusion of the many computer simulations for which there are no counterpart empirical data. In this sense, validation is only a suitable concept for those cases in which a computer simulation is representing an *actual* system, and not a *possible* or *conceivable* system (e.g., a simulation that violates a constant of nature, such as simulation with a gravitational force equal to $G = 1 \, \mathrm{m \, kg^1 \, s^2}$).

Having said this, it is important to mention that, with the introduction of computer simulations in experimental contexts, validation does not exclusively depends on contrasting results against empirical data. Ajelli and team show how it is possible to run different computer simulations and use their results to assert the reliability for each one—in this case, there is not mere convergence of results, but also of key variables (Ajelli et al. 2010).[16]

Figure 4.1 shows in a flow diagram how verification (both code verification and calculation verification) and validation methods are put into effect in standard scientific practice. The *conceptual model* here is the product of analyzing and observing the physical system of interest (i.e., what we called the *scientific model*). In key applications of computational physics (such as computational fluid dynamics, computational solid mechanics, structural dynamics, shock wave physics, and computational chemistry), the conceptual model is dominated by the set of PDEs used for representing physical quantities.

Two further types of model can be identified: a *mathematical model*, from which the computational or simulation model is created, and a *physical model* which, for simplicity, we shall identify with an experiment (recall our treatment of experimentation in Chap. 3). The *computational model*, in our own terminology, is an operational computer program that implements the simulation model as a computer simulation.

---

[14]Whereas this is a valid claim for some form of experimentations in science, it is not so for others such as economics and psychology.

[15]This claim is widely attributed to Edsger Dijkstra.

[16]Strictly speaking, Ajelli et al. are doing *robustness analysis* (Weisberg 2013).

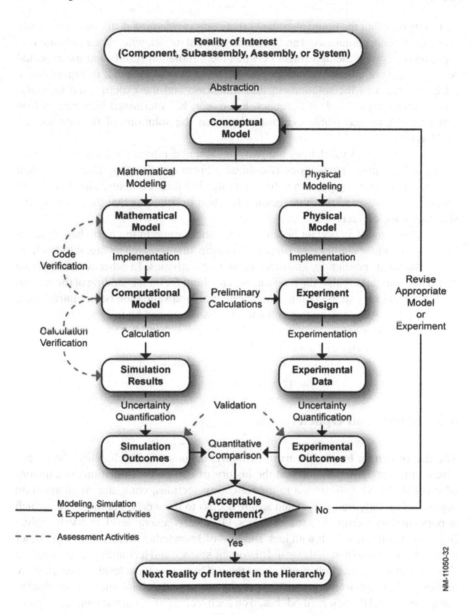

Fig. 4.1 Phases of modeling and simulation, and the role of verification and validation (ASME 2006, 5)

Let us note that the figure also shows that *code verification* deals with the fidelity between the conceptual model and the mathematical model, whereas *calculation verification* deals with the agreement between the simulation results and the expected results of the computational model. *Validation*, on the other hand, is a quantitative adequacy between the solutions of the simulation and the experimental measurements or observations. This adequacy, again, can be determined by a comparison that provides an acceptable agreement between the solutions of the two models involved.[17]

Verifications and validations are two fundamental pillars for asserting the reliability of computer simulations (Durán and Formanek 2018a). This means that researchers are on good grounds for claiming that their computer simulations are reliable processes, and thus that they are justified in believing that the results of the simulation are correct.

This outcome is important because, as we shall argue in the next chapter, reliability of computer simulations support the claim that researchers are able to claim for explanation, prediction, and other epistemic activities. In other words, reliable computer simulations also allow researchers to claim for understanding of the results.

A final and very short comment on verification and validation before introducing the next topic. Although neither software nor hardware can be fully verified or validated, researchers are still developing methods that reduce the occurrence of errors and increases the credibility of the simulation. This interest works as evidence for the importance of both methods for the general reliability of computer simulations.

4.3 Errors and Opacity

The use of certain terms, like trust, certainty, accuracy, and reliability should not give us the wrong impression that the history of computer simulations is a history of success. As we will discuss in the following sections, computer simulations do not provide a fully transparent and secure access to the world, but rather one which is populated with errors and uncertainties. On a very general level, every discipline in human history has dealt with lack and loss of knowledge. In this sense, computer simulations are also part of the grand history of science and technology, and therefore there is nothing particularly novel about them. On a more local level, however, there are several issues that emerge strictly from the use of new technologies—in particular, computers—and the new methods based on such technology—in particular, computer simulations.

Up to now, my efforts were located in looking at computer simulations as sources for knowledge about the world. This view is of course correct, since computer simulations do constitute powerful methods for providing accurate information about the world that surrounds us. In fact, we dedicate the next chapter to discussing how computer simulations explain, predict, and perform several epistemological func-

---

[17]For a more detailed discussion on Fig. 4.1, see Oberkampf and Roy (2010, 30).

tions. But as mentioned earlier, computer simulations are also a source of opacities, errors and uncertainties that might affect their reliability and undermine the trust researchers put on the results of a simulation. Because of this, it is important to discuss, however briefly, the sources of errors and opacities that pervade the practice of computer simulations, as well as the philosophical issues tailored to them. Equally important will be to discuss some of the standard mechanisms used by researchers that help mitigate these errors and circumvent opacities. Let me say that my treatment here will be irremediably unfair given the complexity of these issues. I hope, however, to be able to sketch the basic problems and suggested possible solutions..

4.3.1   Errors

Errors are a time-honored problem in scientific and technological disciplines. While some are an indication that something went wrong, some others can actually offer some insights into what went wrong (Mayo 1996). In any case, the importance of studying errors is that at some point they will affect the reliability of computer simulations, and therefore the researcher's trust on their results. For this reason, researchers need to provide strict and robust measures for detecting errors as well as ways to prevent them. When they occur, however, it is important to know how to deal with them, reduce the negative effects, and revert them when possible.[18]

A first approach to errors divides them into *arbitrary errors*, such as the product of a voltage dip during computation or a careless laboratory member tripping over the power cord of the computer; and *systematic errors*, that is, errors that are inherently to the process of designing, programming and computing a simulation model.

Arbitrary errors have little value to us. Their probability of occurring is very low, and they do not make any contribution to the frequency of a computing process for producing correct results. If anything, they are the silent and lonely contributors that, once detected, are easy to make disappear. It comes as no surprise that researchers put little effort in trying to understand their probability of happening. Rather, research facilities devise protocols and safety measures that help cope with them—if not fully eradicate them. Cables installed in special cable canals away from the main working area is a safety measure that successfully avoids careless laboratory members tripping over them. Likewise, power sources are very stable nowadays, and some facilities even have backup generators. In case of a general blackout, data is usually replicated across several servers in such a way that computer simulations can be resumed from the last execution. Thus, arbitrary errors should not be considered a threat to the general reliability of computer simulations.

---

[18]For an excellent analysis on errors and how they affect scientific practice in general, see Mayo (2010), Mayo and Spanos (2010a). For how errors affect computer science in particular, see Jason (1989). And for the role of errors in computer science, see Parker (2008). I am taking here that errors negatively affect computation.

Systematic errors, on the other hand, play a more prominent role. I divide them into two kinds, namely, *hardware errors* and *software errors*. As its name suggests, the first kind of error relates to the malfunctioning of the physical computer, whereas the second stems from errors in design and programming computer simulations—including errors from out-of-the-box packages.[19]

4.3.1.1 Hardware Errors

Physical errors are related to the temporal or permanent malfunctioning of the computer microprocessor, memory, I/O device errors and, in general, any physical component of the computer that could alter the normal computational process of a simulation.

Of all imaginable sources of hardware errors, perhaps the most eccentric one come from outer space: supernovae, black holes, and other cosmic events can in fact cause computers to crash. These hardware errors are known as 'soft errors' and are produced by a simple astronomical phenomena: cosmic rays hitting the Earth's atmosphere. When cosmic rays collide with air molecules, they produce an 'air shower' of high energy protons, neutrons, and other particles that can hit the computer's internal components. If they get near the wrong part of a chip, the electrons they trail create a digital 1 or 0 out of nowhere (Simonite 2008).

These kind of error are called 'soft errors' because their presence does not result in permanent damage to the computer, nor in changes to the physical characteristics of the hardware. Rather, soft errors only corrupt one or more bits in a program or a data value, altering in this way the data and therefore the computing process, without producing visible damage to the hardware. In the mid 90s, IBM tested nearly 1,000 memory devices at different sea levels—valleys, mountains, and caves—and the result showed that the higher the altitude, the more soft errors would occur, whereas the ratio of soft errors on devices tested in the caves was almost zero.

Soft errors typically affect the computer's memory system as well as some combinational logic used in circuits, such as the Arithmetic Logic Unit. In the case of memory circuits, the source of soft errors are the energetic particles that generate enough free charge to upset the state of the memory cell. In the case of combinational logic, voltage spikes or transient current can upset the clock timing causing soft errors that propagate and which are finally latched at the output of the logic chain (Slayman 2010). The obvious outcome are unknown unreliable results.

The following is a simple example that illustrates how harmful this type of error could actually be. Consider the following two subroutines in the *C Language*:

---

[19]For an overview of errors in the design and production cycle of computational systems, see Seibel (2009); Fresco and Primiero (2013); Floridi et al. (2015).

Algorithm 1 Bitwise operator

---
if (a & b) **then**
   printout "The Democrats have won"
else
   printout "The Republicans have won"
end if

Algorithm 2 Logic operator

---
if (a && b) **then**
   printout "The Democrats have won"
else
   printout "The Republicans have won"
end if

In the *C Language*, operators like & arc called *bitwise* because they are operations performed on a bit level by simply changing a 1 into 0, and vice-versa. In addition, the *C Language* does not include the concept of a Boolean variable. Instead, 'false' is represented by *0*, and 'true' could be represented by any numeric value not equal to *0*. This fact, usually well hidden, allows programmers to write, under certain circumstances: *a & b* just as *a && b*.

Suppose now that a $=$ 4 (00000100_2), and b $=$ 4 (00000100_2). In this case, the conditional in both algorithms are evaluated to be equal and thus 'The Democrats have won' is the final output. This is the case because the logic operator && evaluates to non-zero values (Algorithm 11), and thus the first statement is always called. Similarly, the bitwise operation adds to a non-zero value a & b $= (00000100_2)$ in (Algorithm 10).

Now, if a soft error occurs on the 3rd bite rendering a $= 0$ (00000000_2), then the logic operator in algorithm 11 will still evaluate to "The Democrats have won" simply because it is a non-zero value, whereas the bitwise operator would evaluate to "The Republicans have won" since it evaluates to a zero value a & b $= (00000000_2)$ (Algorithm 10).

The example shows that if the right energetic particle hitting the right place in the memory where the value of 'a' is stored, then the results could be quite different—in this case, with Democracy being the main casualty. Arguably, the example is highly unlikely, just as much as it is unfortunate. The real possibilities of such an error actually happening are extremely small, if not virtually impossible. And this not only because probabilities are against it, but also because programming languages have become very stable and robust. Despite these considerations, it is a real possibility that manufacturers take very seriously. As hardware components reduce in dimensions and power consumption, and as RAM chips become more dense, the sensitivity to radiation increases dramatically, and thus the likelihood of a soft error happening (Baumann 2005).

To counteract this effect, technological companies focus on making better chip designs and improving on error checking technologies. In fact, the computer giant Intel has been systematically working towards incorporating a built-in cosmic ray detector into their chips. The detector would either spot cosmic ray hits on nearby circuits, or directly on the detector itself. When triggered, it activates a series of error-checking circuits that refresh the memory, repeat the most recent procedures, and asks for the last message sent to the affected circuit (Simonite 2008). In this way, Intel aims at reducing soft errors and thus increasing reliability in the hardware component.

There are of course other kinds of systematic hardware errors besides soft errors. These typically come in combination with the software that handles the hardware. Perhaps the most famous—or infamous, I should say—hardware and software error in history is ascribed to the Pentium FDIV bug, of the Intel microprocessor. The goal of Intel was to boost the execution of a floating-point scalar by 3 times and the vector code by 5 times, compared to previous microprocessors. The algorithm used would have a lookup table to calculate the intermediate quotients necessary for the floating-point division. This lookup table would consist of 1066 entires, 5 of which, due to a programming error, were not downloaded in the programmable logic array. When these 5 cells were accessed by the floating point unit, it would fetch a 0 instead of the +2 expected, which was supposed to be contained in the 'missing' cells. This error threw off the calculation and resulted in a less precise number than the correct answer (Halfhill 1995). Despite the fact that the chances of the bug appearing randomly was calculated to be about 1 in 360 billion, the Pentium FDIV bug cost Intel Co. a loss of about \$500 million in revenue with the replacement of the flawed processors.

The lesson to take home is that more advanced technology is not immediately translated into more reliable computations. Soft errors appear with the introduction of the modern circuit board and silicon-based technology. In truth, however, hardware errors are of the least concern to most researchers working on computer simulations. This is mainly because, as we have discussed so far, specifying, programing, and running computer simulations is a software driven practice. Researchers rely on their hardware, and they have very good reasons to do so. Moreover, when the question about the reliability of computer simulations arises, most philosophers, myself included, are thinking of ways to deal with software errors, not with hardware errors. On this note, let us now turn to discuss software errors.

4.3.1.2 Software Errors

Software errors are, arguably, the most frequent source of errors in computer science. They lead to instabilities in the general behavior of the computer software, and seriously compromise the reliability of computer simulations.

Software errors can be found in a myriad of places and practices. For instance, the practice of programing is a chief source of software errors, as programming can get extremely complicated. A faulty compiler and a flawed computer language also bring into focus concerns about the reliability of computer software. Furthermore, *discretization errors* are a major source of errors in computer simulations as they stem from the process of interpolating, differentiating, and integrating a series of mathematical equations.

Surely some of these errors can be avoided, but others prove more complicated. Poor programing, for instance, can be blamed on a clumsy programmer. C. Lawrence Wenham listed several signs that makes a bad programmer (Wenham 2012). These include the inability to reason about the code (e.g., the presence of 'voodoo code', or code that has no effect on the goal of the program but is diligently maintained anyway), poor understanding of the language's programming model (e.g., creating multiple versions of the same algorithm to handle different types or operators), chronically poor knowledge of the platform's features (e.g., re-inventing classes and functions that are built-into the programming languages), the inability to comprehend pointers (e.g., allocating arbitrarily big arrays for variable-length collections), and difficulty seeing through recursion (e.g., thinking that the number of iterations is going to be passed as a parameter). The list extends significantly. Programing, in any case, is an intellectually demanding activity in which even the most skilled and talented programmer is subject to make mistakes.

Let us note in passing that these software errors share the fact that they are all human-related. As pointed out, errors in programing are mostly made by programmers. A silly—and yet disastrous example is the Mars Climate Orbiter with which NASA lost all contact almost a year after its launch in December 1998. The board in charge of investigating the accident concluded that eight factors contributed to the disaster, one of which was the ground-based computer models responsible for the navigation of the probe. A programming error made the computer models fail to translate non-SI units of pound-seconds (i.e., English units) into SI metric newton-seconds (metric units).[20]

At this point one could conclude that software errors are human-tailored and therefore, eradicable with the proper training. One could think, then, that this is true even for cases of a faulty compiler and a flawed computer language, for they are based on bad specifications and missing implementing procedures (e.g., an error call), and therefore are also human-tailored. Furthermore, even the process of discretization of mathematical equations are human-tailored, for they are still carried out for the most part by mathematicians—or computer scientists, or engineers. At the end of the day, software errors are human errors.

Things are in fact a bit more complex than here depicted. There are several sources of errors that do not depend on hard-to-eradicate programming habits—or mistakes produced from our limited cognitive capacity—but from error propagation by the iterative process of computing; that is, the kind of software errors that computers, rather than humans, introduce in the process of simulating. An example can illustrate this point. One way to solve nonlinear functions is by approximating the results with iterative methods. If things go well, that is, if the discretization procedure and the

---

[20] Arthur Stephenson, chairman of the Mars Climate Orbiter Mission Failure Investigation Board, actually believed that this was the main cause of losing contact with the Mars Climate Orbiter probe. See Douglas and Savage (1999).

posterior programing into a simulation model are correct, then the set of solutions of the simulation converge to the right value with a small margin of error.[21] Although this is standard practice, in a handful of cases the set of solutions are inaccurate because of a continuous accumulation of errors during the computation. These kinds of errors are known as *convergence errors*, and they become a matter of worry when their presence goes unnoticed.

It is well known that the two most important convergence errors are *round-off errors* and *truncation errors*. Typically, the former are introduced by the word size of the computer, resulting in the limited accuracy of real numbers. Truncation errors, on the other hand, are errors made by shortening an infinite sum to a smaller size and approximating it by a finite sum.

In order to illustrate the potential negative effects of round-off errors in a computer simulation, consider again the example presented on page 5 of a satellite orbiting around a planet. There, if the equation corresponding to the amount of total energy E (Eq. 1.1) is to decrease, then the semi-major axis a must become smaller. But if the angular momentum H (Eq. 1.2) needs to be constant, then the eccentricity e must become smaller. That is to say, the orbit must round off.[22] Explained in this way, the trend in the orbital eccentricity is steadily downwards, as shown in Fig. 1.3.

Now, it is well known that researchers make use of computer simulations because they are cheaper, faster and easier to set up than to actually construct a satellite and put it into orbit. Many philosophers have even claimed that, because of these reasons, computer simulations act *as if* they were the real satellite orbiting around a planet under tidal stress. I believe that nothing is further from the truth. Researchers are well aware of the limits of their simulations, and know that even if the spikes shown on Fig. 1.3 can be ascribed to a real-world satellite orbiting around a real-world planet and so on, they still cannot ascribe the trend steadily downwards that they see in the same visualization. Why not? Because this is *not* something that actually happens in the real-world, but an artifact of the simulation (i.e., a round-off error). If this effect were actually to be ascribed to the behavior of the real-world satellite, then we would see the orbit of the satellite become circular. But again, this is merely the artifact of computing a round-off error in the simulation model (Durán 2017b). Woolfson and Pert are of course well aware of this fact—as they are also programmers—and therefore can take proper measures to avoid them or, when unavoidable, to deal with such an error.

As an example of measured truncation-errors, we can mention the Runge-Kutta method which, incidentally, is also used in the simulation of the satellite orbiting around the planet. Researchers estimate that this algorithm has a local truncation-error on the order of $\mathcal{O}(h^p + 1)$, and a total accumulated error on the order of $nCh^{p+1} = C(\bar{x} - x_0)h^p$. However, since both are iterative functions, the derivation of the local

---

[21] Provided, naturally, that there are no hardware errors involved.

[22] This is the interpretation of the exchange between energy and angular momentum (Woolfson and Pert 1999a, 18). Let it be noted that the authors do not speak of 'errors', but solely of rounding-off the orbit. This example also shows that rounding-off errors can be interpreted as an inherent part of the programing of a computer simulation. This, of course, does not prevent them from qualifying as 'errors.'

and total error will depend on each iteration and thus it is hard to pin down the exact error. Woolfson and Pert include in the subroutine *NBODY* of their simulation the Runge-Kutta subroutine with automatic step control, and suggest researchers to 'watch' for any unintended result.

We can now easily anticipate that previous knowledge of the existence of an error as well as having means to measure it constitute significant advantages for the overall assessment of results of computer simulations. In this particular simulation, the authors point out that although the simulation of a satellite orbiting around a planet is relatively simple to understand, there are certain effects that might not have been obvious at first sight. The fact that the spikes would occur, for instance, is one such effect. A good simulation, the authors believe, will always bring new, unexpected and important features of the target system under investigation (Woolfson and Pert 1999a, 22). To this thought, we must add that the presence of errors also bring new and unexpected results that researchers must learn how to deal with.

4.3.2 Epistemic Opacity

The previous discussion was an effort to show how errors can contribute to the general inaccuracy of results and thus compromise the reliability of computer simulations. As presented, there are as many sources of errors as there are ways of dealing with them. All things considered, we can justifiably say that many errors—but not all, naturally—are corrigible to a given extent, and therefore they are not so critical to the reliability of computer simulations. Unfortunately, in computational science—and therefore in computer simulations—there is a far more worrying source of mistrust than errors, that is, *epistemic opacity*.

The history of this concept goes back well before the use of computers for scientific purposes. However, it was Paul Humphreys who introduced the term as a distinctive mark of computer science (Humphreys 2004). According to him, an essential feature of epistemic opacity is that researchers are unable, as cognitively limited human beings, to know all the relevant states of a given computational process at any given time. The argument is quite compelling. It says that no human—or group of humans—could possibly examine every element of the computational process relevant for the assessment and justification of the results. Again, 'justification of the results' here simply means to have reasons to believe that the results are correct. Epistemic opacity, then, is conceived as the unrecoverable loss of knowledge about a given computational process followed by the inability to justify the results of such a process (Humphreys 2004, 148). To characterize epistemic opacity more formally, I reproduce Humphrey's definition:

> a process is essentially epistemically opaque to [a cognitive agent] X if and only if it is impossible, given the nature of X, for X to know all of the epistemically relevant elements of the process. (Humphreys 2009, 618)

Let us break up this characterization into its main components. First, the kind of *process* that Humphreys has in mind is a *computational process*—such as computing a simulation model. One could of course ask if there are non-computational processes that also qualify as epistemically opaque. As suggested, the concept is not reserved only for computer processes, but rather it has a long history in mathematics and sociology. Later in this section, I discuss the viewpoint of two mathematicians and a philosopher who claim for three forms of opacity with mathematical and sociological roots that also affect computer simulations.

Another important component in the definition above is the notion of *epistemically relevant elements* for each process. As far as we can tell, an epistemically relevant element in a computational process are any function, variable, memory pointer and, in general, any component directly or indirectly involved in the computation of the model for the purposes of rendering results. Finally, the cognitive agent X refers to any number of researchers involved in an epistemically opaque process. The amount of researchers is, of course, irrelevant.

We can now positively reconstruct Humphreys' characterization of epistemic opacity in the following way. Computer simulations are epistemically opaque to any number of researchers if and only if it is impossible to know the evolution over time of the variables, functions, etc., in the computational process. The consequence of epistemic opacity is, again, that researchers are unable to justify the results of their simulations.

Thus understood, epistemic opacity is a sound argument that puts pressure on computer simulations as novel methods in the sciences and engineering. Indeed, if researchers cannot justify their results, what sort of reasons do they have to trust them and, therefore, to use the results for predictions and explanations? To illustrate the problem, consider again the simulation of the orbiting of a satellite under tidal stress as discussed in Sect. 1.1. If the simulation is stopped at any random stage, it is beyond any number of researchers to reconstruct the current state of the simulation, to retrodict previous states, and to predict future states. Thus, researchers are unable to justify the spikes shown in Fig. 1.3 as the behavior of a real-world satellite under tidal stress. For all they know, the spikes could simply be noise or an artifact of the computation. Epistemic opacity, then, gives many philosophers good reasons to reject the claim that computer simulations are reliable sources of information about the world (e.g., Guala (2002) and Parker (2009)).

To put epistemic opacity in even more perspective, contrast it with some forms of error. As previously discussed, some hardware errors can be reverted and entirely neutralized by having redundancies in the system, for instance. Software errors are, in many cases, anticipated by means of good programming practice and verification and validation methods. When errors are solely considered, then, our lack of knowledge is meant to be only temporary. Once detected and amended, our knowledge of the computer process is restored, and with it the researchers' ability to justify the results of computer simulations. Epistemic opacity, on the other hand, suggests a deep and permanent loss of knowledge, an irreversible uncertainty about a computational process that researchers are not able to control or revert. As a consequence, results are beyond any possible justification.

Epistemic opacity is, therefore, not a form of error. That much is clear. One could reasonably argue, however, that epistemic opacity is familiar to us in a similar way that *abstraction* and *idealizations* are familiar. The argument here is that all three are forms of giving away degrees of detail about a given process (e.g., a target system, a computer process, etc.), and thus a way to lose knowledge. But unlike epistemic opacity, the notions of abstraction and idealization refer to ways of neglecting some aspects of the process in order to enhance our knowledge of it. Researchers abstract from the color of the sand in the Sahara desert because it is absolutely irrelevant for estimating its age (Kroepelin 2006; Schuster 2006). Similarly, idealizations take place in the reconstruction of aerosol's indirect effects on mixed-phase and ice clouds, as they are not included in the simulation used by Benstsen et al. (2013, 689). Contrary to epistemic opacity, then, abstractions and idealizations have the general purpose of enhancing our knowledge, not diminishing it. Furthermore, unlike abstraction and idealization, the presence of epistemic opacity is imposed onto researchers, rather than picked out by them.

The key to understand epistemic opacity is to look at mathematics and how they handle the notion of 'surveyability' of proofs and calculations. Mathematical truths such as theorems, lemmas, proofs, and calculations are in principle surveyable; that is, mathematicians have cognitive access to the equations and formulae, as well as to each step of a proof and calculus. With the introduction of computers, surveyability in mathematics becomes a bit more opaque. An historically interesting example that illustrates the kind of epistemic anxiety that such opacity produces is the proof of the four-color theorem by Kenneth Appel and Wolfgang Haken (Appel and Haken 1976a, b). Donald MacKenzie recalls that when Haken presented the proof, the audience split into two groups roughly at the age of forty. Mathematicians over forty could not be convinced that a computer could deliver a mathematically correct proof; and mathematicians under forty could not be convinced that a proof that took 700 pages of hand calculations could be correct (MacKenzie 2001, 128). The anecdote shows how *surveyavility* is at the center of the epistemological confidence in a mathematical and computational method. In the end, Appel and Hanken had to provide independent reasons for why their program was reliable and therefore rendered trustworthy results.

Under this heading, it is relatively simple to draw connections between surveryability and epistemic opacity: the former prevents the latter. In the era of computers, however, one might rightly ask whether it is necessary to survey a computer simulation at all in order to claim reliability and trust. The purpose of running computer simulations seems to be, precisely, to sidestep complicated calculations by using the machine to do the hard work. In fact, the implications that follow epistemic opacity grossly contrast with the success of computer simulations in scientific and engineering practice.[23] If they are epistemically opaque, and opacity entails lost of knowledge, then how are computer simulations so successful in science and engineering research?

---

[23] For an example of epistemically opaque but successful computer simulations, see Lenhard (2006).

The answer to this question has already been given at the beginning of this section. Reliabilism, as we discussed it earlier, is the most successful way to circumvent epistemic opacity. By the end of this chapter I will show how reliabilism helps in this endeavor. But before, we need to address all conceivable forms of opacity for computer simulations.

In a recent article, Andreas Kaminski, head of the philosophical department at the High Performance Computing Center Stuttgart (HLRS), Michael Resch, director of the HLRS, and Uwe Küster, head of the department of numerical methods presented three different forms of opacity which they called: *social opacity*, *technological opacity*, and *mathematical opacity*, which has an *internal* and an *external* interpretation (Kaminski et al. 2018). All three are forms of opacity concerned with the reliability of computer simulations and the degree to which researchers can trust their results. Let us briefly discuss them in turn.[24]

Social opacity is another name for the division of labor, largely discussed among social epistemologists. When projects are too complex, take long in time, or include a large number of participants, then division of labor is the best road to success. Take for instance measuring a quantity in nature. Physicists typically know how to detect such a quantity, what instrument to use, and how to analyze the data. They might even know within what range such a quantity should be expected, and what it means for a given physical theory. The engineers know little about the physicist's measuring work and concerns. Instead, they know to the minutiae how to construct a precise instrument capable of detecting the quantity of interest. Finally, we have the mathematicians, silent contributors that work out the mathematics for the instrument and, sometimes, for the physical theory as well. This is, of course, a simplified and rather idealized case of division of labor. The point is to illustrate that different researchers collaborate towards the same goal, in this case, to detect and measure a quantity in nature by means of a precise instrument. Division of labor is a highly successful strategy that involves researchers across disciplines—as well as diverse researchers within the same discipline.

Kaminski et al. claim that, in a context of division of labor, researchers know about their own work but not others, and therefore they must rely on the expertise, solutions, and professional standards that are not their own (Kaminski et al. 2018, 267). The example is a computer simulation implementing a module linked to some software library. Typically, such modules and libraries circulate among research projects, different communities, and technicians to the point that no one, on their own, knows all the details of the module. There is a large body of literature in social and technological studies that back up their claim: instruments, computer

---

[24] Another author's ideas on epistemic opacity and epistemic trust worth considering is Julian Newman. To Newman, epistemic opacity is a symptom that modelers have failed to adopt sound practices of software engineering (Newman 2016). Instead, by means of developing the right engineering and social practices, Newman argues, modelers would be able to avoid several forms of epistemic opacity and, ultimately, reject Humphreys' assertion that computers are a superior epistemic authority. As he explicitly puts it: "[...] well architected software is not epistemically opaque: its modular structure will facilitate reduction of initial errors, recognition and correction of those errors that are perpetrated, and later systematic integration of new software components" (Newman 2016, 257).

modules, and artifacts are not only a technological product, but a social one (Longino 1990). Thus understood, *social opacity* is the lack of knowledge that researchers of one community have on a technological product—or a technological change—from another community.

Technological opacity, on the other hand, borrows from early ideas in mathematics where researchers make use of theorems, lemmas, and a host of mathematical machinery without having specific knowledge of the formal proof that grants their truth (Kaminski et al. 2018, 267). The authors' idea is that something similar can be said about technological instruments. Researchers make use of a myriad of instruments regardless of any deep understanding they may or may not have of the instrument. By 'deep understanding,' Kaminski et al. mean any insight that goes beyond simply knowing how to successfully make use of the instrument.

Although social and technological opacity are acknowledged as being relevant sources that negatively affect the assessment of results, the authors put more weight in *mathematical opacity* as the central form of epistemic opacity for computer simulations. In this context, they claim for two forms of mathematical opacity, namely, an *internalist* and an *externalist* form (Kaminski et al. 2018, 267). Internal mathematical opacity is conceived as the cognitive agent being unable to survey the simulation model due to its complexity. Indeed, it is extremely difficult, if not impossible, to survey a simulation model that includes complex mathematical properties (e.g., commutative, distributive, etc.), and computational machinery (e.g., conditionals, I/O communication, etc.). External mathematical opacity, on the other hand, consists in a cognitive agent being unable to solve the mathematical model by her own means, and therefore having to implement it on the computer. It follows that the computing process for solving the model is no longer cognitively accessible by the agent. Thus understood, the externalist approach is very similar to the ideas of epistemic opacity presented by Humphreys.

In this context, two questions come to mind. First, we need to ask to what extend these forms of opacity actually represent a problem for the assessment and justification of the results. Let us recall that to justify the results means to have reasons to believe that the results are correct. Insofar as epistemic opacity is a source of mistrust, the question must be asked. Second is the question of whether there are ways to circumvent any form of epistemic opacity. My answer is that there are. In fact, I have already presented a solution at the beginning of this chapter. Let us now answer each question in turn.

Kaminski et al. are correct in pointing out that the social, the technological, and the internalist viewpoint of mathematics are forms of epistemic concern. I am skeptical, however, that they represent a problem for the assessment of results of computer simulations. My reasons stem from the fact that Kaminski et al. do not make explicit what constitutes an epistemically relevant element for each process. When we make those elements plain, it becomes clear that these forms of opacity do not necessarily jeopardize the justification of results of computer simulations. To put the same point more specifically, I identify two characteristics of these processes that make them more epistemically 'transparent'—however such transparency can be measured—and thus not a real threat for the justification of the results of a computer simulation.

First, all three forms of opacity depend on the right amount of description. Typically, researchers are only interested in a limited amount of information that counts for the justification of results. When the right amount is obtained, then they have the intended level of transparency. For instance, knowing that a pseudo-random engine module produces the following numbers $\{0, 763, 0, 452, 0, 345, 0, 235 \ldots\}$ could be less epistemically relevant for justifying the results of the simulation than knowing that the results fall within $0 < i < 1$. The reason is that researchers might prefer the latter formulation because it is sufficiently accurate, simpler in formulation, and mathematically more manageable. There are no intrinsic reasons that make the knowledge of each pseudo-random number more epistemically relevant than simply providing a range.

The right amount of description, then, is a way to lessen the pressure of social, technological, and mathematical processes being epistemically opaque. As the example of internal mathematical opacity shows, providing a range rather than each individual value contributes better to the justification of results.

With this in mind, one could also elaborate on examples about social and technological opacity. For instance, many researchers have no idea how computers locate variables and their values in the memory. However, this fact does not prevent programmers from specifying in their coding where in the memory a variable must be located. By knowing this, researchers are able to justify why the results have a given truncation error—say, because it is an 8 MB memory and the size of the value stored is of 16 MB. This is an example of how technological opacity does not necessarily affect the justification of results.[25] Furthermore, researchers could justify the results of their simulations without having any information of how the procedures of storing and fetching values in memory were designed and programmed. In other words, social opacity also does not entail lack of justification.

A second characteristic that speaks in favor of epistemic transparency is the *right level of description* of a process. This is the idea that epistemically relevant elements are tailored to the description at different levels in the process. Unlike the previous characteristics that emphasize the amount of information, here the focus is on the depth of a given amount of information. So, at low levels of description some processes are opaque, whereas at some higher levels they are not. In a technological process, for instance, researchers typically do not know to the *minutiae* every epistemically relevant element tailored to the instrument, but this hardly seems to be an argument in favor of opacity. To illustrate this point, imagine a fictitious case where researchers know every detail of the workings of a physical computer, from the part that every transistor plays in the computer, to the physical laws involved that allow the computer to work as it does. Then comes the question: would any researcher benefit from this excess of knowledge or, instead, would it be a burden to the justification of results? It seems rather obvious that a depth knowledge of a process could turn out to be, in fact, counterproductive.

---

[25]Humphreys has used a similar argument to point out that researchers do not need to know the details of an instrument in order to know that the results of such an instrument are correct (e.g., that the observed entity actually exists) (Humphreys 2009, 618).

In the case of social processes, for instance, researchers exchange ideas and relevant information with colleagues about design and programming decisions on the functionality of a software module. Social processes are not obscurantist practices, but rather they are well documented (Latour and Woolgar 1987). This viewpoint also applies to the internalist viewpoint of mathematics, had we believed the claim that mathematics are, to some extent, a social process (De Millo et al. 1979).

Far from establishing the epistemic transparency of social, technological, and internal mathematical processes, the two characteristics discussed above aim at raising concerns about these processes' alleged opacity. What Kaminski et al. call 'opacity' is, in fact, a neglecting stance over these processes. Division of labor consists in neglecting detailed knowledge of others researchers' work in order to move our own forward. Technological processes neglect information about instruments and apparatuses in order to be able to use such technology more efficiently. And finally internal mathematical processes use a similar neglecting principle, since they neglect information over specific steps of a proof in order to facilitate the establishment of further mathematical truths.

In this respect, social, technological, and internal mathematical processes neglect information in order to enhance our epistemic insight. In other words, they are not meant to undermine the justification of the results, but rather to enhance their epistemological assessment. Researchers are familiar with these forms of neglecting as they use them systematically in abstractions and idealizations. Standard philosophy of science takes that *abstraction* aims at ignoring concrete features that the target system possesses in order to focus on their formal set-up (Frigg and Hartmann 2006). *Idealizations*, on the other hand, come in two flavors: while Aristotelian idealizations consist in 'stripping away' properties that we believe not to be relevant for our purposes, Galilean idealizations involve deliberate distortions (Weisberg 2013).

The similarity between all three forms of opacity and abstraction and idealization stems, again, from the fact that all these processes are meant to enhance our trust in the results of a simulation rather than to undermine it. Social processes are exclusively devised for the success of collaboration. Something similar can be said about technological processes. Modularization, for instance, has been created to facilitate researchers' focus on what is most relevant about their work. Progress in science heavily depends on these forms of opacity, just as much as it depends on abstraction and idealization.[26]

For the reasons given above, it seems that we cannot classify social, technological, or the internalist viewpoint of mathematical processes as epistemically opaque in the sense given at the beginning of this section; that is, that our loss of knowledge cannot be reverted, neutralized, or anticipated. This is of course not to say that they do not constitute an epistemological issue in their own right. They do raise

---

[26]Humphreys himself draw parallels between social processes and social epistemology, and concludes that there is no real novelty in either that would affect computer simulations to a greater extent than they affect any other scientific, artistic, or engineering discipline (Humphreys 2009, 619).

important questions regarding scientific and engineering practice, but in principle there is nothing related to the problem of epistemic opacity that interests us here.

External mathematical opacity, or simply epistemic opacity, is a rather different animal. Whereas social, technological, and internal mathematical opacity pose humans at the center of their analysis, in the context of Humphreys' epistemic opacity—or Kaminski et al.'s external mathematical opacity—humans do not have such a relevant role. Instead, Humphreys focuses on the process of *computing* and in how it becomes epistemically opaque. Thus understood, computer processes, and not humans, are key for understanding epistemic opacity. This is the reason why Humphreys claims that humans have been displaced by computers from the center of production of knowledge. Humans, to rephrase Humphreys, are part of an old epistemology.

We could now answer our second question, which aims at considering the ways to circumvent epistemic opacity.[27] Interestingly, the answer to this question can be traced back to the beginning of this chapter, where we discuss forms of granting reliability to computer simulations.[28]

To recap briefly, recall from Sect. 4.4 our discussion on how researchers are justified in believing that the results of a computer simulation are correct of a target system. There, we said that there is a reliable process—the computer simulation—whose probability that the next set of results is correct is greater than the probability that the next set of results is correct given that the first results were just luckily produced by an unreliable process. A computer simulation is a reliable processes because there are well established verification and validation methods that confer trust on the results.[29] In other words, our confidence in the results of an epistemically opaque process such as a computer simulation is given by processes external to the simulation itself, but that ground their reliability—and which are, in and by themselves, not opaque.

4.4 Concluding Remarks

Building trust in computer simulations and their results is not a simple matter. To some, there is an insurmountable epistemological barrier imposed by the very nature of the computer and computer processes that will never allow us, humans, to know how the process of simulation is taking place. Such a viewpoint enables the claim that computer simulations are not as reliable as laboratory experimentation, and thus their epistemological significance must be reduced—some of this was already dis-

---

[27] In fact, reliabilism might be used to circumvent all forms of opacity (v.gr. social opacity, technological opacity, and internal mathematical opacity).

[28] I introduce and discuss reliabilism in the context of computer simulations for the first time in Durán (2014).

[29] In Durán and Formanek (2018a), we extend the sources of reliability to a history of (un)successful computer simulations, robustness analysis, and the role of the expert in sanctioning computer simulations.

cussed in Chap. 3. To others, myself included, we do not need to have full epistemic transparency on a computational processes in order to claim knowledge. Rather, researchers can genuinely know something about the world regardless of the opaqueness involved in the simulation. One, of course, still needs some minimum criteria of what constitutes a reliable computer simulation in order to have any claims on knowledge. One such criteria is secured by entrenching computer simulations as reliable processes.

Throughout this chapter, the aim was simply to show the many discussions surrounding epistemic trust in computer simulations, as well as the many philosophical avenues that these discussions must transit. Sharpening some concepts helped us to better grasp the underlying problems, but unfortunately it is never enough. Here, I took a clear position, one that conceives that trust in computer simulations can be granted on epistemic grounds, and that reliabilism is the way to go. The next chapter assumes much of what has been said so far by showing how such trust is displayed in the scientific and engineering arena. I called them 'epistemic functions' as a way to highlight the multiple sources of understanding offered by computer simulations in scientific and engineering practice.

Chapter 5
Epistemic Functions of Computer Simulations

The previous chapter made a distinction between knowing and understanding. In computer simulations, this distinction allows us to set apart knowing when researchers trust the results and when they understand them. In this chapter, we explore different forms of understanding by means of using computer simulations. To this end, I have divided the chapter between epistemic functions that have a *linguistic form*, from those that are characterized for having a *non-linguistic form*. This distinction is meant to better help categorize the different ways in which researchers obtain understanding of the world that surround us by means of using computer simulations. Indeed, sometimes computer simulations open up the world to us in the form of symbols (e.g., by the use of mathematics, computer code, logic, numerical representation), whereas sometimes the world is accessed through visualizations and sounds. In the following, I analyze studies on *scientific explanation*, *predictions*, and *exploratory strategies* as linguistic forms that provide understanding of the world, and *visualization* as a case for non-linguistic forms.

5.1 Linguistic Forms of Understanding

5.1.1 Explanatory Force

Any theory of scientific explanation aims at answering the question '*why q?*',[1] where q could be virtually any proposition. Consider the following *why*-questions: 'why did the window break?', 'why does the number of students that drop out of school increase

---

[1] In some cases, we can expect an explanation by asking '*how q*' questions. For instance, 'how did the cat climb the tree?' is a question that demands an explanation of how the cat managed to get up a tree. Although some theories of explanation put *how*-questions at their core, here we will only be interested in *why*-questions.

© Springer Nature Switzerland AG 2018
J. M. Durán, *Computer Simulations in Science and Engineering*,
The Frontiers Collection, https://doi.org/10.1007/978-3-319-90882-3_5

every year?', 'why is $\frac{1}{x}$ undefined for $x = 0$' in the context of classical infinitesimal calculus? Researchers answer these questions in different ways. Take for instance the first question. A researcher could correctly explain a broken window by pointing out that a rock thrown at it caused the window to break. Another researcher could have an explanation that appeals to the hardness of the materials—minerals and glass—and the fact that the former causes the latter to break. Because the minerals that compose the rock are harder than those that compose the glass, the window will break every time a rock is thrown at it. Yet another researcher could use as an explanation the molecular structure of the materials, and show how the properties of one structure causes the other to break. Regardless of the level of detail used for the explanation, they all point to the fact that it is the rock that *causes* the window to break.

Appealing to causes for providing an explanation is not always attainable or even suitable. Consider our third question of why is $\frac{1}{x}$ undefined for $x = 0$? No amount of causes can actually provide a good explanation for this *why*-question. Instead, we must derive the answer from a set of schemata using the theory of calculus. One such explanation is the following: consider $\frac{1}{x}$ as x approaches to 0. Now consider its positive and negative limit. Thus, $\lim_{x \to 0^+} \frac{1}{x} = +\infty$, whereas $\lim_{x \to 0^-} \frac{1}{x} = -\infty$. It follows that $\lim_{x \to 0} \frac{1}{x}$ does not exist, and therefore is undefined.

The above explanations are meant to illustrate two basic components in any theory of explanation. First, the explanatory relations that allow us to answer *why*-questions can take several forms. Sometimes researchers can explain by pointing out the *causes* that bring q about, whereas sometimes it is better to *derive* q from a corpus of scientific belief—like scientific theories, laws, and models.[2]

The basic requirement for the causal approach is that we identify how q fits into the causal *nexus*. That is, which are the causes that bring about q as its effect. So, the rock (i.e., the cause) made the window break (i.e., the effect). The chief challenge for any causal approach is to spell out the notion of a *cause*, a truly difficult matter. Here is where important philosophical work is needed.

The alternative to a causal approach consists in deriving q from a set of well-established beliefs. We have already shown how this could be done in the case of explaining why $\frac{1}{x}$ is undefined for $x = 0$. Interestingly, advocates of this viewpoint of explanation also claim that their account could explain cases such as the broken window. To this end, researchers need to reconstruct as schematic sentences a few well-known theories relevant for the explanation and derive the fact that the window broke. Some obvious candidates are Newtonian mechanics for the trajectory of the

---

[2]The two major theories of explanation are *ontic* theories, where causality is at the center and *epistemic* theories, where derivation is at the center. The main advocates for the former are Salmon (1984), Woodward (2003), and Craver (2007). The main advocates for the latter are Hempel (1965), Friedman (1974), and Kitcher (1989). The reader interested in the many other theories of explanation should approach (Salmon 1989).

rock, chemistry for the chemical bonding characteristics of glasses, and material theory for specifying the type of glass as well as its physical properties, among others.

The second component in any theory of explanation is *understanding*. Why should researchers ask *why*-questions? What reasons do scientists and engineers have to be interested in explaining such and such? The answer is that by explaining, researchers are able to advance their understand of why something is the case. Explaining why the rock broke the window advances our understanding of physics (e.g., trajectory of projectiles), and chemistry (e.g., the resistance of crystals), as well as the simple phenomenon of a window broken by a rock.

To put the issue of understanding into some perspective, consider how explanation is fundamental in rejecting false theories. An interesting historical example is the *phlogiston theory*, the predominant view in the late eighteenth century on chemical combustion. According to this theory, combustible bodies are rich in a substance called phlogiston that is released into the air upon burning. Thus, when wood is burned, phlogiston is emitted into the air leaving ash as a residue. Now, the phlogiston theory stands upon two principles that hold upon combustion, namely, that the burned body loses mass, and that air is 'filled' with this substance called 'phlogiston.'

The phlogiston theory came under pressure when it was unable to explain how, upon combustion, certain metals actually gained mass instead of losing it, violating the first principle. In an attempt to save the theory, some proponents suggested that the phlogiston actually had negative mass, and therefore instead of lightening the total mass of the body, it would make it heavier agreeing in this way with most measurements. Unfortunately, such a suggestion begs the question of what it means for phlogiston to have negative mass, a concept that was not accountable by the physics of the late eighteenth century. Other proponents suggested that the phlogiston emitted by these metals was, in fact, lighter than air. However, a detailed analysis based on Archimedes' principle showed that the densities of magnesium along with its combustion could not account for a total increase in mass. Today we know that the phlogiston theory was unable to explain the increase of weight in some metals upon burning, making it a false theory about combustion.

The importance of scientific explanation for computer simulations is twofold. On the one hand, it furnishes simulations with a chief epistemological function, namely, to provide understanding of what is simulated. On the other hand, it strips away the simple role of simulations as finding the set of solutions to an unsolvable mathematical model—the standard approach in the problem-solving viewpoint. In this context, there are three questions that interest us here. These are, in order: 'what do we explain when we explain with computer simulations?', 'how is explanation for computer simulations possible at all?', and finally 'what sort of understanding should we expect by explaining?'

Answering the first question used to be rather straightforward matter: researchers want to explain real-world phenomena. Such is the standard format found in most theories of explanation. Either a theory, a hypothesis, or a model among many other units of explanation have the explanatory force to account for a phenomenon in the world.

This is the idea put forward by Ulrich Krohs (2008) and Paul Weirich (2011), and questioned later by myself (Durán 2017b). To these authors, the explanatory power of computer simulations stems from the underlying mathematical model that is implemented on the computer capable of accounting for real-world phenomena.[3] Weirich makes this point clear when he asserts that "[f]or the simulation to be explanatory, the model has to be explanatory" (Weirich 2011, Abstract). Similarly, Krohs argues that "in the triangle of real-world process, theoretical model, and simulation, explanation of the real-world process by simulation involves a detour via the theoretical model" (Krohs 2008, 284). Although these authors disagree in their interpretation of how mathematical models are implemented as a computer simulation, and in how they represent the target system, they agree that simulations are merely instrumental devices for finding the set of solutions to mathematical models. Thus understood, the mathematical models implemented in the simulation have the explanatory force rather than the computer simulation itself.

My viewpoint differs from Krohs and Weirich in that, to me, researchers have access first and foremost to the results of the simulation, and thus their interest in explanation lie in accounting for such results. Naturally, researchers will be eventually interested in also understanding the real-world that is represented by their results. However, such understanding of the world comes at a later stage. To illustrate my position, consider the example of the spikes in Fig. 1.3. Researchers have access to the spikes in the visualization, and that is what they want to explain. The question to them is, then, 'why do the spikes occur' and 'why is there a steadily downwards trend'? The importance of explaining the results of the simulation is that researchers are in the position to also explain the real spikes, had a real satellite, planet, distance, tidal force, etc. as specified in the simulation model been out there in space. Thus, by running a reliable computer simulation and explaining its results, researchers can explain why certain phenomena in the real-world occur as shown in the simulation, without actually engaging into any interaction with the world itself. Explanation in computer simulations is a crucial feature that makes plain their epistemic power, regardless of comparisons with scientific models or experimentation. Furthermore, as I will show later in this section, the only unit capable of accounting for the results is the simulation model, as opposed to the mathematical model that Krohs and Weirich claim.

To sum up, Krohs and Weirich take that mathematical models implemented in the computer simulation have explanatory force, while I argue that it is the simulation model the unit of analysis that should actually holds that role. Furthermore, Krohs and Weirich believe that explanation is of a real-world phenomenon, whereas I claim that researchers are first interested in explaining the results of the simulation, and later the real-world phenomenon that they represent.

The next question is how is scientific explanation for computer simulations possible at all? To answer this question we must refer back to the beginning of this section. There, I mentioned two main approaches to scientific explanation, namely, the causal

---

[3]Recall our discussion on computer simulations as problem-solving techniques in Sect. 1.1.1.

approach that requires one to locate q in the causal *nexus*, and the inferential account that entrenches explanation by derivation of q from a set of scientific beliefs.

To illustrate this coarse-grained terminology, let us again use the example of the satellite under tidal stress and explain why the spikes in Fig. 1.3 occur. The causal explanation consists in showing that, as an initial condition, the position of the satellite is at its furthest distance from the planet, hence the spikes only occur when they are at their closest. When this happens, the satellite is stretched, which is caused by the tidal force exerted by the planet. Correspondingly, inertia causes the satellite's tidal bulge to lag behind the radius vector. The lag and lead in the tidal bulge of the satellite give spin angular momentum on approach, and subtract it on recession. When receding from the near point, the tidal bulge is ahead of the radius vector and the effect is therefore reversed. The spikes are then caused by the exchange between spin and orbital angular momentum around closest approach (see Woolfson and Pert (1999a, 21)). The inferential account, instead, would first reconstruct the simulation model as schematic sentences—implementing, among other things, Newtonian mechanics—and then show how a derivation of the spikes as shown in the visualization takes place.

Although both explanations seem to be valid, there is a fundamental difference that sets them apart. Whereas in the *causal* approach the explanatory relation depends on an objective external relation (i.e., causal relations), in the *inferential* approach the explanation is quantified over the set of current scientific knowledge and established beliefs. Because computer simulations are abstract entities, much like mathematics and logic, it is rather natural to think that an explanation of the spikes depends on a corpus of scientific beliefs (i.e., the computer simulation) rather than on exogenous causal relations.[4] In Durán (2017b) I argue in favor of the first position.[5]

At this point one might be tempted to suppose that, if the simulation model explains the results and the results are the byproduct of reckoning the simulation model, then there must be some sort of argumentative circularity between what researchers want to explain (i.e., results of the computer simulation) and the explanatory unit (i.e., the simulation model)? To address this issue, let us recall the distinction between *knowing* and *understanding* introduced earlier. It is important that we do not confuse what researchers know about the results of the simulation, that is, that the simulation model reckons the model and renders results, with what researcher understand from the results. Researchers explain because they want to understand something. Explaining results of a simulation help them understand why some result happened, regardless of the knowledge of how and that it happened. Take once more the example of why the spikes in Fig. 1.3 occur. The fact that researchers know that the results are correct of the model says nothing of why the spikes are occurring. Unless researchers carry out an explanation explicitly saying that the spikes occur due to an exchange

---

[4]In Sect. 6.2 I briefly discuss attempts to claim for causal relations *in* computer simulations, that is, whether researchers would be able to infer causal relations from computer simulations. This issue should not be confused with implementing a causal model, which is perfectly possible given the correct specification.

[5]Another important issue that speaks in favor of explanation by derivation is that results of computer simulations carry errors that causal theories are unable to account for (see Durán (2017b)).

between spin and orbital angular momentum around closest approach, they have no real clue of why those spikes are there. Explanations work, when they do, not only in virtue of the right explanatory relation, but also because they provide genuine scientific understanding. The Ptolemaic model, for instance, could not explain the trajectory of planets in any epistemically meaningful way since it fails to provide understanding of the planetary mechanics. On the other hand, classical Newtonian models explain precisely because they describe in a comprehensible way the structure of planetary motion. The criterion for entrenching one type of model rather than the other as explanatory lies, partially, in their capacity for yielding understanding of the phenomenon under scrutiny.

We finally come to our last question, that is, 'what sort of understanding should we expect by explaining with computer simulations?' As we already mentioned, researchers want to explain because they expect to gain further understanding of the phenomenon under scrutiny and, in doing so, to make the world a more transparent and comprehensible place.

It is a well known fact in philosophy that our understanding can take different forms (Lipton 2001). One such form is to identify understanding with having good reasons to believe that something is the case. Under this interpretation, explanations provide good reasons to believe the results of computer simulations—or to believe that the world behaves in the way computer simulations describes it. Although this viewpoint is attractive, it fails to differentiate between knowing that something is the case from understanding why it happens. The fact that simulations show that there should be many more small galaxies around the Milky Way than are observed through telescopes provides an excellent reason to believe that this is in fact the case, but not the slightest idea as to why (Boehm et al. 2014).

Another way explanation provides understanding is by reducing the unknown to something more familiar—and therefore known. This view is inspired by examples such as the kinetic theory of gases, where unfamiliar phenomena are compared to more familiar phenomena such as the movement of tiny billiard balls. Unfortunately, this view suffers many drawbacks, including, issues about the meaning of 'being familiar.' What is 'familiar' to a physicists might not be familiar to an engineer. Additionally, many scientific explanations relate familiar phenomena to unfamiliar theory. There is perhaps nothing more familiar to us than a morning traffic jam on our way to work. However, events such as this require very complex modeling where explanation is far from familiar.

Let us note that the familiarity viewpoint also has difficulty with the so-called 'why regress.' This means that only what is familiar is understood, and that only what is familiar can explain. It follows that this viewpoint does not allow that which is not itself understood can nevertheless explain. But researchers are interested in allowing this: they want to be able to explain phenomena even in cases where they do not understand the theories and models involved in the explanation.

A perhaps more sophisticated way of understanding consists in pointing out the causes that bring a given phenomenon about. This is the form that takes most causal accounts of explanation. Our first question of why the window broke can be fully understood when we account for all the causes that lead to that scenario or, as philosophers of science like to say, the phenomenon is located in the causal *nexus*. Thus, when a rock is thrown at the window, then there are a series of causal relations that eventually lead to an effect: the rock thrown by my hand travels through the air and eventually reaches out the window that finally breaks.

To my mind, none of these forms of understanding are suitable for computer simulations. In some cases, computer simulations do not give researchers any sort of reasons for believing their results. In others, reduction to the familiar is simply impossible if one takes into account the large amount of uncertainties specified and unspecified in the simulation model. It is ludicrous to think that some sort of reductionism from more complex simulations to less complex ones—and perhaps more familiar—is even possible. Finally, the possibility of identifying causal relations and pointing them out a simulation seems far fetched. As I discuss later in Chap. 6, causality in computer simulations is rather an open research program than an initial assumption.

Besides the previous interpretations of understanding, there is yet another form that turns out to be quite promising for computer simulations. This interpretation is known as the 'unificationist' viewpoint because it takes that understanding consists in seeing how, what has been explained, fits into a unified whole.

To the unificationist, understanding comes from seeing connections and common patterns in what initially appeared to be brute or independent facts.[6] 'Seeing' here is taken as the cognitive maneuver of reducing the explained results—or real-world phenomena—to a greater theoretical framework, such as our corpus of scientific beliefs. Several philosophers of science have ascribed to this viewpoint of understanding—although not necessarily to unificationism. Gerhard Schurz and Karel Lambert say that "to understand a phenomenon P is to know how P fits into one's background knowledge" (Schurz and Lambert 1994, 66), and Catherine Elgin asserts that "understanding is primarily a cognitive relation to a fairly comprehensive, coherent body of information" (Elgin 2007, 35). The reduction proposed by the unificationist comes with several epistemological advantages, such as results becoming more transparent to researchers, who in turn obtain a more unified picture of nature, as well as strengthen and systematize our corpus of scientific beliefs. Overall, says the unificationist, the world becomes a more simplified place (Friedman 1974; Kitcher 1981, 1989).

In Durán (2017b) I argue that when results of a computer simulation are understood by means of explaining them, a similar cognitive maneuver is performed. Researchers are able to incorporate the results of the simulation into a larger theoretical framework, reducing in this way the number of independent results seeking an explanation. Thus, by explaining why the spikes in Fig. 1.3 occur, researchers are broadening their body of scientific knowledge by incorporating a case derived from Newtonian mechanics.

The case of computer simulations is particularly interesting for it is actually carried out in two steps. First, the results are included into the body of scientific beliefs related to the simulation model; and second, they are included into our greater body of scientific beliefs.

---

[6]For an analysis of 'brute' and 'independent' facts, see Barnes (1994) and Fahrbach (2005).

Let me illustrate this last point with an explanation of the occurrence of the spikes in Fig. 1.3. By explaining why the spikes show up in the visualization, researchers are giving reasons for their formation. Such explanation is possible, I believe, because there is a well-defined pattern structure that enables researchers to derive a description of the spikes from the simulation model (Durán 2017b). Understanding the spikes, then, comes from incorporating them into the larger corpus of scientific beliefs that is the simulation model. That is, researchers grasp how the results fit into, contribute to, and are justified by reference to the theoretical framework postulated by the simulation model. This is precisely the reason why researchers are able to explain the occurrence of the spikes as well as their downwards trend: both can be theoretically unified by the simulation model. Furthermore, since the simulation model is dependent on well-established scientific knowledge—in this case Newtonian mechanics—researchers are able to see the results of the simulation in a way that is now well-known to them, that is, unified with the general body of established scientific beliefs relating two-body mechanics.

Thus far the standard image of the unificationist applies to computer simulations. But I believe that we can extend this image by showing how understanding results also encompass a *practical* dimension. From the perspective of simulation research, understanding results also involves grasping the technical difficulties behind programing more complex, faster, and more realistic simulations, interpreting verification and validation processes, and conveying information relevant for the internal mechanism of the simulation. In other words, understanding results also feeds back into the simulation model, helping to improve computer simulations. For instance, by explaining and understanding the reasons why the spikes trend downwards, researchers are aware of the existence of and have the means to solve roundoff errors, discretization error, grid resolution, etc. In computer simulation studies, researchers want to explain because they also want to understand and improve their simulations, just as much as they want to understand real-world phenomena (Durán 2017b).

The last point that we need to address here is how to understand real-world phenomena by means of explaining results of computer simulations. As I introduced at the beginning of this section, Weirich and Krohs had as their chief purpose the explanation of real-world phenomena using computer simulations. Both authors are right in thinking that the use of computer simulations is justified, in a large number of cases, because they provide understanding of certain aspects of the world. The question now is, then, can we understand real-word phenomena by explaining results of computer simulations? I believe that we can answer this question positively. We know that the visualization of the results of the simulation represent the behavior of a real-world satellite under tidal stress. This means that the results of the simulation related to the spikes represent, and thus can be ascribed to, the behavior of a real-world satellite. In this sense, and following Elgin on this point (2007; 2009), there is an entitlement—via representation and ascription—of the results of the simulation to the behavior of a real-world satellite. It is precisely because of this entitlement that we are able to relate our understanding of the results of the simulation with our understanding of the behavior of the real-world satellite. This point can also be made by means of the practical ability that presupposes understanding something. As Elgin cogently argues, the understander holds the ability to make use of the information at her disposal for practical purposes (Elgin 2007, 35). In our case, researchers could actually build the satellite specified in the simulation and put it up in space.

5.1.2 Predictive Tools

When philosophers focused their attention on scientific explanation, they also turned to scientific prediction. In fact, Carl Hempel and Paul Oppenheim, the two main philosophers to systematize and set the agenda on philosophical studies on scientific explanation, believed that prediction was the other side of the same coin. These ideas flourished in and around 1948 with their seminal work *Studies in the Logic of Explanation* (Hempel and Oppenheim 1948), and continued until the demise of logical empiricism in 1969 at a symposium on the structure of theories in Urbana, Illinois (Suppe 1977).

Despite the joint birth, scientific explanation and prediction have pretty much gone in different directions ever since. While studies in scientific explanation grew significantly establishing different schools of thought, philosophers working on prediction are more difficult to find. It is interesting to note that when it comes to philosophical studies on computer simulations, much more effort has been put into studying prediction and much less into scientific explanation. This asymmetry can be explained by the practices of computer simulations in scientific and engineering contexts. Researchers are more interested in predicting future states of a system, rather than in explaining why such states are obtained. At the beginning of this chapter, we have discussed at some length scientific explanation. Now is the time to draw attention to some of the basic principles of scientific prediction in the context of computer simulations. But first, an example that illustrates predictions in science and helps to introduce the basic terminology.

A beautiful case in the history of science is Edmond Halley's predictions of comets. The general beliefs about comets at the time of Halley was that they were mysterious astronomical interlopers moving unpredictably through the sky. Although Halley made some precise retrodictions—that is, predictions into the past—establishing that the comet that appeared in 1531, 1607, and 1682 were all manifestations of the same phenomena, his postdictions—that is, predictions into the future—of the next appearance of the comet were not as successful, confirming in this way the established scientific belief of the time. In fact, he postdicted that the comet would show up in the sky again by 1758, one year earlier than its actual appearance.

It was the job of Alexis Clairaut, a prominent Newtonian, to correctly postdict the next appearence of the comet, which was set to reach its perihelion by 1759. Clairaut based his postdictions on calculations that would include forces unknown at the time, but that made sense within the Newtonian theory. Such forces mainly refered to the actions and influence of distant planets—recall that Uranus was discovered in 1781 and Neptune remained unknown until 1846. It is interesting to note that, besides a successful postdiction, what was also at stake was the confirmation of Newtonian theory as the most adequate way to describe the natural world. The struggle among factions lead many physicists to at first reject Clairaut's calculations, and many others to anticipate with some joy the failure of Newtonian theory. The story ends with Clairaut's calculations correctly postdicting the next appearance of Halley's Comet—despite some truncations in the higher terms of his equation—and with the Newtonian view of the world imposing overwhelmingly over less adequate theories.

The example of Halley's Comet makes plain the importance of prediction for scientific research—and, in this particular case, also for confirming the Newtonian theory. Successful predictions are valuable because they go beyond what is know most directly and evidently by researchers, providing 'hidden' information about phenomena and empirical systems.[7]

Two important characteristics about prediction are the *temporal* dimension and the *accuracy* of the prediction. A common mistake is to presume that predictions are about saying something meaningful about the future (i.e., postdictions), but not about the *past* (i.e., retrodictions). This is typically the case when retrodictions are mistakenly confounded with scientific evidence of something happening. Thus, it is said that Halley found evidence that the comet that appeared in 1531, 1607, and 1682 were the same, but not that he made retrodictions. Although evidence and prediction can, in some cases, be related, they are still two separate notions. Retrodictions—as well as postdictions—are part of a theory's implications, and this is so regardless of temporal constraints. Evidence, on the other hand, serves to either support or counter a scientific theory. In this sense, retrodictions and postdictions are the manifestation of the same phenomenon, namely, of *predictions*. Thus understood, it is correct to say that Halley predicted the three occurrences of the comet prior to his observation of 1682, just as Clairaut predicted its next appearance in 1759 because he made calculations using a theory—a form of theory implication.

The language of prediction is used to describe declarative assertions about the past as well as future events made in light of a theory, and thus we shall use it here. In fact, the temporal dimension entails an epistemic component: "to predict is to make a claim about matters that are not already known, not necessarily about events that have not yet transpired" (Barrett and Stanford 2006, 586). In other words, prediction is about the 'diction' of an event, past, present, or future. Thus understood, prediction is asking 'what does the theory tell us?', 'what knowledge is new?' In our case, the answer is rather obvious: the next appearance of Halley's Comet.

---

[7]Incidentally, these same features render prediction inherently risky, since access to hidden information typically depends on shared standards in a given community, and is therefore potentially intersubjective. Now, one way to avoid problems of intersubjectivity is by demanding convergent predictions. That is, by means of having disparate areas of research all predicting towards similar results, our confidence in the prediction must inevitably increase. This is what philosopher Heather Douglas calls *convergent objectivity* (Douglas 2009, 120). We must keep in mind, however, that there are other forms of objectivity as well. In here, we are not going to worry about questions on subjectivity and objectivity, as they deserve a study of their own. For further references, the reader is referred to Daston and Galison (2007), Lloyd (1995), and of course (Douglas 2009).

Closely related to this temporal dimension is the question of 'how well does the theory predict the actual observed outcome?' This is a chief question that gives us a purchase on the second characteristic about prediction relevant for our discussion, namely, the accuracy of predictions.

In the previous chapter we discussed accuracy as the set of measurements that provide an estimated value close to the true value of the quantity being measured. For instance, Halley's Comet's nucleus is about 15 km long, 8 km wide, and roughly 8 km thick—a rather small nucleus for the vast size of its coma. The comet's mass is also relatively low, an estimated 2.2×10^{14} kg. Astronomers calculated the average density to 0.6 g/cm^3 indicating that it is made of a large number of small pieces held loosely together. With this information at hand, along with calculations of the trajectory, astronomers can accurately predict that Halley's Comet is visible to the naked eye with a period of 76 years.

The accuracy of scientific predictions depend on a combination of the nature of the event, the adequacy of our theories, and the current state of our technology. Although it is true that by using computers we are able to predict with higher degrees of accuracy than Clairaut's the next time Halley's Comet will show up in the sky,[8] predictions are possible because the trajectory of the comet can adequately be described using Newtonian mechanics.

Now, for many occasions it is difficult to obtain *exact* predictions. This is so because of the nature of the phenomena to predict. Indeed, there are many events and phenomena for which we can only predict their behavior within a given range of probability of occurrence. A simple example is the following. Suppose I ask you to chose a card and place it face down on the table. Then I have to predict its suit (i.e., 'hearts', 'diamonds', 'clubs', or 'spades'). Theory says that I have a 1 in 4 chance of being right. The theory does not entail which suit you just placed face down, but rather that the probability of me being correct is $\frac{1}{4}$. Although this does not strictly count as a prediction, it tell us something about the accuracy of the prediction.

Another good example of inexact predictions are so-called chaotic systems. These are systems that are very sensitive to initial conditions, where small errors in computation can rapidly propagate into large predictive errors. This means that certain predictions are impossible after a given point in the calculation. The fact that chaotic systems past a certain point make predictions difficult and essentially impossible in a practical sense has important consequences for scientific research. A standard example of a chaotic system is atmospheric conditions. In weather forecasting, it is very difficult to do long-range predictions due to the inherent complexity of the system which only allows predictions up to a certain point. This is the reason why

---

[8]Clairaut was involved in several computations of the orbital period of the Halley's Comet. For one of the first calculations, he divided the cometary orbit into three parts. First, from 0° to 90° of eccentric anomaly, the first quadrant of the ellipse from perihelion, which the comet took more than 7 years to traverse. Second, from 90° to 270°, the superior half of the orbit, which the comet needed more than 60 years to traverse. Third, from 270° to 360°, the final quadrant of the ellipse, which took about 7 years to traverse. Clairaut's calculations were rectified in different moments throughout history (Wilson 1993).

the behavior of the weather is considered chaotic, the smallest changes in the initial conditions make long-term predictions unfeasible—if not utterly impossible.

Let us now discuss how prediction works in computer simulations by discussing two simulations of an epidemic outbreak in Italy. The first simulation is a stochastic agent-based model, whereas the second simulation is a structured meta-population model—known as the GLobal Epidemic and Mobility—GLEaM. The agent-based model includes an explicit representation of the Italian population through highly detailed data of the socio-demographic structure. In addition, and for determining the probability of commuting from municipality to municipality, the model makes use of a general gravity model standard in transportation theory. The epidemic transmission dynamics is based on an influenza-like illnesses (ILI) compartmentalization based on stochastic models that integrate susceptible, latent, asymptomatic infections, and symptomatic infections (Ajelli et al. 2010, 5). Marco Ajelli and his team, responsible for designing and programing these simulations, define the agent-based model as "a stochastic, spatially-explicit, discrete-time, simulation model where the agents represent human individuals [...] One of the key features of the model is the characterization of the network of contacts among individuals based on a realistic model of the socio-demographic structure of the Italian population" (Ajelli et al. 2010, 4).

On the other hand, the GLEaM simulation integrates high-resolution population databases—estimating the population with a resolution given by cells of 15×15 min of arc—with the air transportation infrastructure and short-range mobility patterns. Many standard GLEaM simulations consist of three data layers. A first layer, where the population and mobility allows the partition of the world into geographical regions. This partition defines a second layer, the subpopulation network, where the inter-connection represents the fluxes of individuals via transportation infrastructures and general mobility patterns. Finally, and superimposed onto this second layer, is the epidemic layer, that defines inside each subpopulation the disease dynamic (Balcan et al. 2009). In the study, the GLEaM also represents a grid-like partition where each cell is assigned the closest airport. The subpopulation network uses geographic census data, and the mobility layers obtain data from different databases, including the International Air Transport Association database consisting in a list of airports worldwide connected by direct flights.

As expected, there are advantages and disadvantages of using the agent-based and the GLEaM simulations. As far as the GLEaM model is concerned, the detailed space mobility networks provide an accurate description of the channels of transportation available for spreading the disease. However, accurate estimations of the impact of the disease at a more local level are difficult to obtain due to the low level detail contained in this model. As for the agent-based approach, although it is highly detailed in regards to the structures of households, schools, hospitals, etc., it suffers from gathering high confidence datasets from most regions of the world (Ajelli et al. 2010, 12). Despite each simulation offering different features, and thus affecting their respective predictive attack rate of the disease, Ajelli and his team notice that there is a steady convergence in the results that provides confidence in each one of the simulations' prediction. The heterogeneity of the transportation network provided by the GLEaM model, for instance, makes possible accurate predictions of the spatio-temporal spread of the ILI disease at a global scale. On the other hand, the explicit representation of individuals in the agent-based model facilitates accurate predictions of the spread of an epidemic on a more local scale.

Getting more specific about the predictions, the difference in the peak amplitudes varies depending on several factors, mainly based on the values for the reproductive number.[9] The average epidemic size predicted by GLEaM is 36% for a reproductive ratio of $R_0 = 1.5$, whereas for the agent-based model with the same reproductive ratio, the average epidemic size goes down to 25%. On the other end of the epidemic outbreak, the epidemic size predicted by GLEaM is 56% of the population for a reproductive ratio of $R_0 = 2.3$, while it is 40% for the agent-based. Researchers observed an absolute difference of about 10% for $R_0 = 1.5$ and of about 7% for $R_0 = 2.3$. Predictions were also made for a reproductive ratio of $R_0 = 1.9$, showing similar behavior in the average epidemic size predicted by both simulations.

To Ajelli et al. these predictions look quite accurate as they see convergence in the average epidemic size. The team, however, is unable to evaluate which of the two predictions is better. The high level of realism of the agent-based model should, in principle, speak in favor of the accuracy of the prediction. But unfortunately such a realistic model is not free of modeling assumptions. Indeed, in order to determine the probability of commuting from municipality to municipality, Ajelli et al. implement a general gravity model used in transportation theory as well as assuming a power law functional form for the distance—despite the fact that other functional forms such as exponential decay can also be considered (Ajelli et al. 2010, 3). Another assumption stems from taking homogeneous mixing in households, schools, and workplaces, while random contacts in the general population are assumed to depend explicitly on distance (Ajelli et al. 2010, 3). Despite the fact that both assumptions are perfectly reasonable, they inevitably affect the prediction and, therefore, the grounds for preferring the agent-based model prediction over the GLEaM's. Similarly, the prediction of the structured metapopulation model is equally hard to asses precisely because of its lack of realism. "The correct value", says Ajelli et al. "should be in between the prediction of the models, as supported by the fact that the difference between the models decreases as R_0 increases, with the models converging to the same value for the attack rate" (Ajelli et al. 2010, 8).

At this point we must highlight a crucial difference between the predictions offered by Ajelli et al. simulations, and Clairaut's: whereas the predictions of the later can be empirically confirmed, the predictions of the computer simulations could not. On

---

[9]In epidemiology, the reproductive ratio—or reproductive number—represents the number of emerging cases that one infected individual generates on average over the course of an infectious period in an otherwise uninfected population. Thus, for $R_0 < 1$, an infectious outbreak will die out in the long run, whereas for a $R_0 > 1$ an infection will be able to spread out and infect the population. The larger the number, the harder it will be to control the epidemic. Thus, the metric helps to determine the speed that an infectious disease can spread through a healthy population. In the case of both simulations, the authors report that for large R_0, the local epidemics become more widespread across all the layers of the population, making the population structure less and less relevant.

what grounds, then, do Ajelli and his team claim that these are accurate predictions? Their answer is to insist that the good agreement reached by the results of both simulations constitute reasons that speak in favor of an accurate prediction.[10] Surely, the fact that both simulations are, in their own terms, good representations of the target system also contributes to the researcher's confidence. But it is convergence of results that firmly grounds the confidence in accurate predictions.

Such a standpoint is not lacking good reasons. Convergence of results does speak in favor of the simulations as well as of the accuracy of the prediction. In the previous chapter, we saw how verification and validation methods can be used to grant reliability on the simulation and thus help researchers to trust in their results. The interesting twist introduced by Ajelli and team is that validation is not against empirical data, but rather against another computer simulation.[11] This means that the results of one simulation works as a confirmatory instance of the results of the other simulation—and vice versa. This is a fairly common practice among researchers making intensive use of computer simulations and from whose target systems there are no empirical data available for validation. In other words, the reliability of computer simulations are used for providing grounds for trust in their results which, under conditions of convergence, serve in turn as grounds for believing in the accuracy of all sorts of predictions.

Having said this, predictive practices should be tempered with some healthy precautionary principles. As any researcher knows, models are based on our representation of how the real-world works. Although the corpus of scientific and engineering beliefs is extremely successful, it is by no means infallible. That is why validation against empirical data is a good and a necessary practice in scientific and engineering research. It virtually forces researchers to 'test' their results against the world, and to find a workaround in cases of a mismatch. In this particular case, Ajelli and his team take convergence of results as a positive sign—although not a confirmatory instance—of an accurate prediction. Thus much is, I believe, correct. However, as they also explain in their article, prediction by convergence of results presupposes a common basis in the modeling frameworks within which the simulations are located. That is, information such as parametrization, data integration, initial and boundary conditions, and even the many approximations used need to be shared amongst researchers, otherwise, Ajelli et al. argue, researchers would be unable to discount unwanted effects such as those stemming from modeling assumptions.

---

[10]This answer begs the question of what is 'good agreement' for results of computer simulations. Ajelli et al. do not specify how this notion should be understood. We could speculate that we have a good agreement when all results fall within a given distribution (e.g., a normal distribution). This, of course, requires specification. As we have learned from studies in the epistemology of experiment, agreement on the results of *different* techniques give confidence not only in the results, but also in the ability of the techniques to produce valid results (Franklin 1986, Chap. 6). The question is, then, to what extent could we consider the agent-based simulation and GLEaM two different techniques.

[11]By this I mean that the results of either computer simulation cannot, in principle, be empirically validated. Since each computer simulation implements sub-models, there is the possibility that some of them have been empirically validated.

More worryingly is the possibility that the results of both simulations are artificially converging. In order to avoid this, researchers try to compare results that lack a common basis among simulations. That is, computer simulations need to be differentiated in terms of model assumptions, initial conditions, target systems, parametrization, calibration, etc. This is the approach chosen by M. Elizabeth Halloran and her team, as reported by Ajelli et. al. Halloran compares the results of three individual-based models of a pandemic influenza strain with different assumptions and initial data, one at the description level of a city and two at the description level of a country (Halloran et al. 2008).

Despite any number of precautionary efforts, comparing different models and results is generally a difficult task. Ajelli et al. point out that Halloran et al.'s comparison is constrained to each model's assumptions as well as to the available simulated scenarios. In this respect, they never explicitly define a common set of parameters, initial conditions, and approximations shared by all models. The low transmission scenario proposed by Halloran, for instance, is compared in each model by using different values for the reproductive number, with the risk of not being able to discount the effect of this difference in the obtained results. As a consequence, convergence of results is artificial and so is any prediction made by the simulations.

5.1.3 Exploratory Strategies

To us, exploratory strategies have the character of a research activity with the aim of generating significant findings about phenomena without having to appeal to, nor relying on the theory of such phenomena. Exploratory strategies, then, has two goals. On the one hand, it intends to bring about the observable changes in the world; on the other, it serves as a testing ground for new, yet to be stabilized concepts.[12]

The first question that comes to mind is, then, why is it important to set apart theory from the experiment? One obvious answer consists in pointing out that any given theory might not be capable of providing all the relevant information about a given phenomenon. A good example of this is Brownian motion. In 1827, the botanist Robert Brown noticed that particles trapped in cavities inside pollen grains in water would move. Neither Brown nor anyone at the time was able to determine nor explain the mechanisms that caused such a motion. It was not until 1905 that Albert Einstein published a paper explaining in precise detail how the motion that Brown had observed was the result of the pollen being moved by individual water molecules.

Another important reason for decoupling theory from experiment is that, in many cases, experiments are used to debunk a given theory. For instance, August Weismann

---

[12]Exploratory strategies is one of those topics that attract little attention from philosophers, despite their centrality in the scientific and engineering endeavor. Fortunately, there are three excellent pieces of work that cover these issues. Friedrich Steinle on exploratory experiments (Steinle 1997), Axel Gelfert on exploratory models, (Gelfert 2016) and Viola Schiaffonati on exploratory computer simulations (Schiaffonati 2016). Here, I will take a different approach from these authors.

conducted an experiment where he removed the tails of 68 white mice repeatedly over five generations just to show that no mice were born without a tail or even with a shorter tail simply because he cut them. Weismann's interest was to put an end to Lamarckism and the theory of inheritance of acquired characteristics by showing how it could not account for a generation of mice with short or no tails.

Having proposed the importance of independent experimentation for our general knowledge of the world, can we now give sense to the idea of exploratory strategies in laboratory and computer simulations? To answer this question, we need to go back a bit, both in time and in number of pages, and briefly review the context wherein the ideas about exploratory strategies flourished. In Chap. 3, I mentioned the *logical empiricists* as a group of philosophers and scientists interested in understanding the notion and implications of scientific theories. As they soon discovered, this is as much a familiar term as it is an evasive concept. In particular, they discover that theory and experiment are more interwoven than initially thought.

One core position of the logical empiricist was to regard experiments not so much as a philosophical problem in itself, but as a subsidiary methodology for understanding theories. Critics took advantage of this position to attack the core ideals of the logical empiricists. To many of these critics, experiments and experimentation were issues of genuine philosophical value and had to be addressed as such. One particularly interesting point brought up was the objectivity of observational evidence, and how this issue relates to theory. The problem can be framed as follows: in their experiments, researchers interact, observe, and manipulate real-world phenomena as well as collect evidence for further analysis, just as Weismann, Lamarck, and Brown did in their own way many years ago. The problem then, is to determine whether the observations are objective,[13] or they depend on the researcher's theoretical background knowledge—a problem known as 'theory ladenness.'

Positions on this point were divided. In many cases, researchers could not warrant that experimentation provides significant findings about phenomena without having to appeal in some way to theory. The reason for this is that every researcher approaches the world with some background knowledge. Even Brown, when he observed the particles moving through the water, was approaching the phenomenon with the mind of a scientist. Several critics of the logical empiricists, including Thomas Kuhn, Norwood Hanson, and Paul Feyerabend among others, were highly suspicious of ideas of objectivity in observational evidence. To them, researchers cannot really observe, collect, and use laboratory evidence without committing themselves to a given theory.

To illustrate the extent of the problem, let us take a specific case from the history of astronomy. Early observational astronomers had very simple instruments for observing the stars. One of the first important observations made by Galileo Galilei—besides the number of moons orbiting Jupiter—was of Saturn in the year 1610. Back then, he incorrectly guessed that Saturn was a large planet with a moon on each side. Over the next 50 years, astronomers continued to draw Saturn with the two moons, or with "arms" coming out of the poles. It was not until 1959 that Christiaan Huygens

---

[13]This notion is used in the sense of independent of any researcher, instrument, method, or theory.

correctly deduced that the "moons" and "arms" were actually the ring system of Saturn. This was of course possible due to improved telescope optics. The point is that until the actual discovery of the ring system, astronomers were looking at Saturn in the same way as Galileo depicted it.[14]

The example shows several issues related to exploratory strategies and the problem of theory ladenness. It shows that observation is not always the most reliable source of knowledge simply because instruments might not be powerful enough—or are being manipulated—to actually provide reliable information about the world. It also shows that the researcher's expectations are a major source of influence in their reports. This is especially true for cases where an 'authority' has established the working grounds on a given issue. Galileo is an example of how authority sometimes goes unquestioned.

A combination of these two problems is furnished from the history of physics. In the early 1920s there was a major controversy between Ernest Rutherford and Hans Pettersson concerning the emission of protons from elements such as carbon and silicon undergoing bombardment by alpha particles. Both researchers conducted similar experiments where they were able to observe a scintillation screen for faint flashes produced by particle strikes. Whereas Pettersson's laboratory reported a positive observation, Rutherford's reported not seeing any of the expected flashes from carbon or silicon. James Chadwick, Rutherford's colleague, visited Pettersson's laboratory to evaluate their data in order to work on possible errors in their own approach. While Pettersson's assistants were showing him the results, Chadwick was manipulating the equipment without anybody being aware of this. Chadwick's manipulations altered the normal operating conditions of the instrument, ensuring that no particles could actually hit the screen. Despite this, Pettersson's assistants still reported seeing flashes at very close to the rate reported in previous conditions. After these events, Pettersson's data were uncontroversially discredited (Stuewer 1985, 284–288).

More to our interests, the example shows that the researcher's observations are shaped by their training and by the theory in which they—and their supervisor and associates—frame their experiments. This raises the following question that is at the core of our study on exploratory strategies. If observation—and other forms of experimentation—is tinted by theoretical expectations, in what sense could the whole process of experimentation—from setting up the experiment, to assessing the data, included the manipulation of instruments and phenomena—generate findings about phenomena without appealing to theory?

To be able to answer this question, we need to make the notion of 'theory ladenness' more precise.[15] Friedrich Steinle, a philosopher dedicated to studying many of these issues, has argued that a conception of theory ladenness such as the one expounded here falls short in capturing the complexity and diversity of scientific experimentation (Steinle 1997, 2002). To his mind, we need to discriminate between

---

[14]See Brewer and Lambert (2001) and Van Helden (1974).

[15]The reader must be aware that there are many subtleties involved in the literature on 'theory ladenness' that we are not going to address here. For a good source of discussion, see Hanson (1958), and Kuhn (1962).

two types of experimentation, namely, 'theory driven' experiments, and 'exploratory' experiments. In his view, theory driven experiments exhibit more or less the same features described for observational experimentation in Chap. 3. That is, experiments are set up and carried out with "a well-formed theory in mind, from the very-first idea, via the specific design and the execution, to the evaluation" (Steinle 1997, 69). Now, to say that an experiment is 'theory driven' suggests at least three different meanings. It could mean that the expectations regarding its results fall within the framework provided by such a theory; it could mean that the design of the experiment depends, more or less, on the theory; and it could mean that the instruments used for the experiment are highly theory dependent. Thus understood, theory driven experiments serve several specific purposes, such as the determination of parameters and the use of theories as heuristic tools for the search of new effects.

By contrast, exploratory experiments makes use of strategies characterized by the lack of theoretical guidance. To be more precise, according to Steinle, none of the aforementioned meanings attached to theory driven experiments applies to exploratory experiments. Thus, an exploratory experiment generates findings about phenomena that neither appeal to the framework that the theory provides, nor to the theory used to design the experiment, nor to the theory built in the instruments that are used. In other words, the experiment and its results render relevant information about phenomena on their own account.

One serious problem with this viewpoint is that there is a lack of sophisticated notion of theory in place, as well as a lack of understanding of the levels of theory involved in the design, execution, and analysis of experimental results that could help characterize these strategies accordingly.[16] In fact, Pío García and Marisa Velasco point out that, in order to defend any of Steinle's interpretations, we must first be able to account for the different levels of theory involved in an experiment,[17] as well as to determine in which of these levels theoretical guidance is most relevant (García and Velasco 2013). In sum, there is no one theory guiding an experiment, and it is not clear which set of theories involved in an experiment makes it theory driven.

In this context, the idea that experiments could generate findings about phenomena strictly without appealing to theory begins to look difficult to ground. We could, however, settle with a more general, and perhaps weaker interpretation of what exploratory strategies are, and the context in which they apply. We could, then, characterize exploratory strategies by their relative independence from strong theoretical restrictions and their capacity to generate significant findings that cannot be framed—or easily framed—within a well-established theoretical framework. A paradigmatic example can be found in the early history of static electricity phenomena, performed by Charles Dufay, André-Marie Ampère, and Michael Faraday. As Koray Karaca points out, these experiments were carried out in a 'new research field'

---

[16]These ideas can be found in García and Velasco (2013, 106).

[17]Ian Hacking offers a first approach to the different types and leves of theory involved in an experiment in Hacking (1992).

that, at the time, had neither a well-defined nor a well-established theoretical framework (Karaca 2013). The results, as recorded, help to advance electromagnetism as the discipline we know today.

Thus understood, exploratory strategies are meant to fulfill very specific epistemic functions. They are particularly important for cases where a given scientific field is open to revision due to, say, its empirical inadequacy. For such cases, exploratory strategies play a fundamental role in the fortune of theories since their findings are, by definition, not framed within the theory under scrutiny. They are also important when, weakly framed within a theory, exploratory strategies provide substantial information about the world that is not implied by the theory itself. More generally, the findings obtained by exploratory strategies are significant with respect to a variety of goals, ranging from more practical matters such as learning how to manipulate phenomena, to theoretical goals such as developing an alternative conceptual framework.[18] Steinle also highlights as a major epistemic function of exploratory strategies the fact that their findings might have significant implications on our understanding of existing theoretical concepts. This is the case when, in the attempt to formulate the regularities suggested by exploratory strategies, researchers are required to revisit existing concepts and categories, and are forced to formulate new ones in order to ensure a stable and general formulation of the experimental results (Steinle 2002, 419).[19] In the face of it, there is not a total division between experiment and theory, but rather a more complicated coexistence based on degrees of independence, capacity to produce findings, and the like.

It is interesting to note that, even in current times when technology has advanced so much into the scientific laboratory, the exploratory strategies applied in experimental contexts are still paramount for the general advancement of science and engineering. Furthermore, it would be a mistake to take from these examples that they are bound to historical periods of science, fields of research, or scientific traditions. The work of Karaca on High-Energy Particle Physics, for instance, is a proof of this Karaca (2013).

We must now ask, is it possible to make sense of exploratory strategies for computer simulations? The answer to this question is yes, and it comes in the form of one of the most appreciated uses of computer simulations, namely, their capacity to show us a world to which we cannot easily have access. A standard use of computer simulations consists in investigating how certain real-world phenomena simulated in the computer would behave under certain specific conditions. By doing this, researchers are able to foster their understanding about that phenomena regardless of the theory or model within which the phenomena is framed. In other words, the simulation provides information about phenomena that goes beyond the model which is implemented. For such cases, computer simulations are typically used to produce results for the particular case at hand, and not to deduce or derive general solutions. Let us illustrate this with an example.

---

[18]This idea is discussed by Kenneth Waters in Waters (2007).

[19]More exploratory functions, such as starting points of scientific research, potential explanations, and other functions are discussed in (Gelfert 2016, 2018).

Take the use of computer simulations in medicine. One important case is to investigate the resistance of human bones, for which it is essential to understand their internal architecture. In real material experiments, force is exerted mechanically, measured, and data is collected. The problem with this approach is that it does not allow the researchers to distinguish between the strength of the material from the strength of its structure. Additionally, this mechanical process destroys the bone, making it difficult to see and analyze how the detailed internal structure responds to increasing force. The best way to obtain trustworthy information about the resistance of the human bones is to run a computer simulation.

Two different types of simulations were outlined at the Orthopaedic Biomechanics Laboratory at the University of California, Berkeley, by Tony Keaveny and his team (Keaveny et al. 1994; Nieburc et al. 2000). The first type, a real cow hipbone was converted into a computerized image, a process that involved cutting very thin slices of the bone sample, and preparing them in a way that allowed the complicated bone structure to stand out clearly from the non-bone spaces. Each slice, afterwards, was turned into a digital image (Beck et al. 1997). These digitalized images were later reassembled in the computer creating a high-quality 3-D image of a particular real cow hipbone. The benefit of this simulation is that it retains a high degree of verisimilitude in structure and appearance for each particular bone sample. In this respect, little is added on, removed, filtered, or replaced in the process of preparing the bone and in the process of turning it into a computer model.

In the second simulation, a stylized bone is computerized as a 3-D grid image. Each individual square within the grid is given assorted widths based on average measurements of internal strut widths from real cow bones and angled in relation to each other by a random assignment process (Morgan 2003). The advantages of the stylized bone come in terms of familiarity with the process of modeling. Researchers begin by hypothesizing a simple grid structure to which details and features are added as needed. In this way, an idealized and simplified abstract structure of the bone is created from the beginning.

Thus understood, the first simulation resembles procedures that are closer to experimental setups whereas the second simulation resembles methods found in mathematical modeling practices. In this respect, all the features of the second simulation are chosen by, and thus known to, the researchers. This is not necessarily so for the first simulation, where researchers deal with a material object that still has the capacity to surprise and confound the researchers (Morgan 2003, 223).

In both cases, nevertheless, the simulation consists of the implementation of a mathematical model using the laws of mechanics. The computer then calculates the effect of the force on individual elements in each grid, and assembles the individual effects into an overall measurement of the strength given in the bone structure. It is interesting to note that the simulation also allows a visual display of how the internal bone structure behaves under pressure, as well as the point of fracture.

Both simulations, therefore, strike as exploratory in design and aim: they investigate how the structure of bones behave under specific conditions of stress and pressure, and thus foster the researcher's understanding. Furthermore, both simulations allow researchers to learn about how the architecture of bones respond in real accidents, which are the conditions for a bone breakage, and how bones can best be repaired, all information that extends beyond the mathematical model implemented.[20]

One could, of course, object that these computer simulations, like all other computer simulations, are theory driven in the sense that the mathematical model with the computation constitute the theoretical framework. However, as I have just mentioned, the idea of computer simulations as exploratory experiments is to generate significant findings about phenomena without holding strong ties with theory. Let us recall that a theory driven experiment could have three different meanings, none of which, I believe, is applicable to the sort of computer simulation here exemplified. It could not mean that the expectations regarding the results of the simulation fall within the framework provided by a given theory, since the results provide information that are not contained in the mathematical model. It could also not mean that the design of the experiment depends, more or less, on the theory. The example shows very clearly that, other than implementing a handful of mechanistic laws, there is little theory backing up the simulation. Finally, it could not mean that the instruments used for the experiment are highly theory dependent. Although it is true that the computer is bound by theory and technology, in principle these play little role in rendering reliable results.[21]

The example then shows two simulations used for exploratory purposes. Similarly to what we said about exploratory experiments, these simulations generate a significant amount of finding that cannot (easily) be framed within a well-established theoretical framework. The fact that both implement mathematical models using the laws of mechanics does not help in accommodating the new evidence. The opposite, I believe, is true. That is, the new evidence obtained from the results contribute to the consolidation, reformulation, and revisiting of the concepts, principles, and assumptions of a medical theory about bones, a physical theory about the resistance of materials, and the built-in models in the simulation. These are, according to Steinle Steinle (2002), the chief characteristics of exploratory experiments. In this sense, the results of the simulations are relatively independent from strong theoretical restrictions included in the simulation models.

From this point of view, computer simulations fulfill a role of exploratory strategies in a similar sense to experimentation. Naturally, in the assessment of their results, researchers still need to take note of the fact that it comes from a model, as opposed

---

[20]There is the claim that the information that a computer simulation can provide is already contained in the models implemented. I find this claim particularly misleading for two reasons. First, because there are several cases where computer simulations produce emergent phenomena that was not strictly contained in the models implemented. Thus, the claim misinforms about the scopes of models as well as misrepresents the role of computer simulations. Second, because even if the implemented models contain all the information that the simulation is capable of offering, this fact says nothing about the knowledge that the researchers have. It is virtually impossible and pragmatically senseless to know the set of all solutions of a simulation model. Precisely for those cases, we have computer simulations.

[21]Recall our discussion in Chap. 4 about the reliability of computer simulations.

to interacting more or less directly with an un-theorized world. Other than this, it seems that exploratory activities are deeply involved in the use and epistemic functions provided by computer simulations.[22]

5.2 Non-linguistic Forms of Understanding

5.2.1 Visualization

Explanation, prediction and exploratory strategies amount to provide understanding of the word by means of some form of linguistic expression. In the case of explanation this is quite straightforward. Researchers reconstruct the simulation model, a logic-mathematical expression, in order to offer an explanation of their results. Prediction, on the other hand, is about producing results that could quantitatively tell researchers something meaningful about a target system. Finally, exploratory strategies consist in generating significant findings about phenomena in the form of data (e.g., numbers, matrices, vectors, etc.). Ultimately, all three epistemic functions as presented here are related to forms of linguistic representations. There is also an alternative, non-linguistic mode of representation, that offers important ways of understanding a target system: the visualization of results of computer simulations.[23]

Since visualizations are genuine forms of insight into a target system, they cannot be approached as conveying redundant information already contained in the results of a computer simulation. The philosophical study on visualizations resist any interpretation that reduces them to mere aesthetic contemplation or as a supportive vehicle of more relevant information. Instead, visualizations are taken to be epistemically valuable for computer simulations in their own right. In this respect, visualizations are an integral part of the researcher's arguments, and are subjected to the same principles of strength and soundness, acceptance and rejection as theories and models.

Now, before anything can be said about visualizations, we must first notice that there are several ways to analyze the term. In here, we are not interested in the theoretically convoluted process of post-processing the results of a simulation into a visualization. This means that we are not interested in transformations (e.g., geometrical, topological, etc.), nor in the type of algorithms that operate according to the data (e.g., scalar algorithms, vector algorithms, tensor algorithms, etc.). We are also not interested in any post-processing stance for refining the results. To us, visualizations in themselves are the main issue. We are interested in the visual outcome of a

---

[22]Admittedly under this interpretation, many computer simulations become exploratory. I do not see this as a problem, for this characteristic fits well with the nature and uses of computer simulations. However, I do see philosophers of science objecting to my interpretation, primarily because it strips away the special status that is originally given to some kinds of experiments.

[23]Because we are solely interested in visualization in computer simulations, other kinds of visualizations, such as graphs, photographs, film videos, X-Rays, and MRI images, are excluded from our study.

computer simulation used for epistemological assessment. In short, we are interested in the kind of understanding that is obtained by visualizing a simulation, and what researchers can do with it.

Here, we are also restricting our interest to four different levels of analysis of visualization. Those are: the spatial dimension (i.e., 2D and 3D visualizations), the time-evolution dimension (static and dynamic visualizations), the manipulability dimension (i.e., cases where researchers can intervene in the visualization and modify it by adding information to it, changing the point of observation, etc.) and the coding dimension (i.e., standardizations used for coding the visualizations, like color, position, etc.). Each dimension individually, as well as in association with each others, offer different kinds of visualizations.

The overall significance of visualizations is that they are a complex of spatial distributions and relations of objects, shapes, time, color, and dynamics. Laura Perini explains that visualizations are partly conventional interpretations in the sense that the relation between a picture and the content is not determined either by intrinsic features of the picture or by a resemblance relation between the picture and its reference. Rather, something extrinsic to the picture and its reference is necessary to determine their relation of reference (Perini 2006). A good example of such conventions is the use of some forms of coding, such as color—brightness, contrast, etc.—glyphs, and the like. Let us take a deeper look at color. *Blue* always represents cold in a simulation including temperature. It would infringe upon unspoken rules of scientific practice to change it to *red* or *green*. Similarly for some symbols: an arrow, for instance, represents going—moving, shifting—ahead—up, towards—when its apex is pointing up. Since the interpretation and comprehension of visualizations are often too natural and automatic for researchers, changes would introduce confusion and unnecessary delays in scientific practice.[24] In this sense, the use and application of the right conventions is necessary in order to comprehend visual representations (Perini 2006, 864).

Consider now a case of changing the spatial distribution of an otherwise standard picture. Talking with a group of colleagues at the University of Colorado, at Boulder, about flipping a world map upside down is an example taken from personal experience. In this case, South America would be 'above' and North America would be 'below.' 'North' and 'South' would of course be shifted too. Flipping the map in this way could be a political statement, since we typically attribute 'above'—in the spatial distribution on the map—to be somehow better or more important. My colleagues found the idea quite appealing, since it provokes a 'mental cramp' to see the world 'upside down.'[25] By this I simply mean that it would take us a few

---

[24]Perini explores this idea in Perini (2004, 2005).

[25]Two points to make here. First, the idea of a 'mental cramp' comes from Wittgenstein, who said that philosophical problems are compared to a mental cramp to be relieved or a knot in our thinking to be untied (Wittgenstein 1976). Second, there is not an 'upside down' world, since it is merely the way we—as humans—decided to represent it. As long as North and South are kept fixed, we are able to represent the globe the way we like—a good example of this is the logo of the United Nations. For an artistic representation of this, see the work of José Torres García, *América Invertida*, 1943.

seconds to understand the new spatial distribution of the continent. As mentioned, the standardization of colors, symbols, and notation are fundamental conventional interpretations that facilitate the free flow of scientific practice.

In this context, computer visualizations can fulfill several epistemic functions. Think how much a researcher could understand a target system by simply looking at the relation of the objects distributed spatially and temporally, their visual properties and behavior, colors, and the like. They all contribute to the overall meaning of what the researcher is observing. Moreover, visualizations facilitate the identification of problems in the set of equations that constitute the computer simulation. In other words, a visualization can show where something has gone wrong, unexpected, or simply shows a false assumption in the model. This is of course not to say that visualizations are able to provide technical solutions. Compare this idea with verification and validation methods. In the latter cases, these methods are meant to locate a set of problems (e.g., wrong mathematical derivations in the discretization process) and to provide the theoretical tools that lead to a solution. In the case of visualization, the identification of errors is by visual inspection, and thus it depends on the trained eye of the researchers. In this sense, visualizations are very useful for helping to decide on different courses of actions and providing grounds for informed decisions, but they do not offer formal tools for addressing issues in the simulation model. In this respect, visualizations are also important because they extend the uses of computer simulations into the social and the policymaking arena.

Let me now illustrate this rather abstract discussion with an example of a tornado. One of the main problems with storm science is that the amount of information researchers have from a real tornado is rather limited. Even with highly sophisticated atmospheric satellites images and tornadoes chasers, the current state of scientific instrumentation cannot provide researchers a complete picture of what is going on. For these reasons and others such as the convenience and safety of studying tornadoes from a desktop, researchers are more inclined to study these natural phenomena with the aid of computer simulations. In this respect, Lou Wicker from the NOAA National Severe Storms Laboratory says "[f]rom the field, we can't figure out completely what's going on, but we think the computer model is a reasonable approximation of what's going on, and with the model we can capture the entire story" (Barker 2004).

The simulation that Lou Wicker, Robert Wilhelmson, Leigh Orf and others recreate is the genesis of a supertwister similar to one seen in Manchester, South Dakota in June 2003. The simulation is then based on a model of a storm that includes equations of motion for air and water substances (e.g., droplets, rain, ice), and grid sizes ranging from a five meter uniform resolution to a much higher resolution. Data is used for seeding the pre-tornado conditions, such as wind speed, atmospheric pressure, and humidity near Manchester at the time. Since data is quite sparse, it could represent points separated by distances from 20 m up to 3 km. This means that researchers need to take such a wide range in the spatial distribution into account when analyzing the visualization.

Naturally, the visualization of the tornado is scaled down. The spatial scale of the tornado ranges from a few kilometers high, to a few 100 km wide and deep. The total runtime of the simulation might also vary, depending on the resolution of the tornado and the number of elements included in it—for instance, whether there is more than one tornado, or it is also simulating the destruction of an entire city. Besides, the time-scale of a tornado also depends on its initial formation, evolution, and demise. For these reasons, computer simulations are typically measured in 'storm hours.' The visualization of an F3 type tornado within a supercell in Wilhelmson et al. (2005) represents an hour of storm evolution, although the visualization only takes about one and half minutes.

Let us also note that the integration and choreography of the visualization is as important as the visualization itself. Choices need to be made in order to focus on the most significant data and events. In the visualization of the tornado, thousands of computed trajectories were edited down to a few of the most meaningful ones. There are typically two reasons for editing visualizations in this way: either there is data that is irrelevant for any visualization, or there is data that is irrelevant for certain purposes in the visualization. For instance, Trish Barker reports that a complete visualization of Wilhelmson et al. (2005) would "look like a plate of angel hair pasta" (Barker 2004) due to the overwhelming amount of information to be displayed. In such cases, the visualization fails in its purpose, as it would provide little understanding of the phenomenon at hand—a similar point is raised with scaling down the visualization in time and space.

In most cases, however, researchers do make use of all the available data, they just highlight different aspects of it. In this way, it is possible to obtain distinct information from the same simulation. And distinct information leads to different courses of action, prevention measures, and the identification of problems in the simulation that were not anticipated during the design and programming stages. An interesting example of the latter is given by Orf, who mentions that in their visualization of the genesis of the tornado, rain was not centrifuging out of the tornado as it should. He concludes: "that's something we need to work on" (Orf et al. 2014).

The goal of the simulation, then, is not to merely compute complex mathematical models, but rather to visualize the structure, formation, evolution, and demise of large, damaging tornadoes produced in supercells. To achieve this, a crucial aspect of the visualization is its realism. That is, the visualization must have sufficient resolution to capture low-level tornado inflow, thin precipitation shafts that form 'hook echoes' adjacent to the main tornado, clouds and, if possible, the ground level over which the tornado passes and all the debris it scatters.

Visual realism is more than an aesthetic feature, it is essential to the assessment of computer simulations. Orf recalls asking tornado chasers their opinion on the visualization of his simulation (see the images in Figs. 5.1 and 5.2). To the field experts, the visualization looks fairly reasonable, which is to say, the visual realism is convincing despite lacking a deep understanding of the simulation model and the methods for visualizing the tornado. To the tornado chasers, there is only the aesthetic experience that approaches the real experience. But this does not diminish their appreciation for the epistemological power of computer simulations. Reliability of the simulation model and trust in the results are provided by Orf and his team, as they are responsible for this in the division of labor.

Another important element in visual realism is color and glyph coding. As Robert Patterson narrates, colored stream tubes represent the motion and velocity of particles when released into the airflow, showing the airflow geometry in and around the tornado.[26] The variation in color of the stream tubes convey additional information of the air temperature, and the cooling down-heating up process—stream tubes are orange when rising and light blue when falling. Interestingly, color plays a further informative role, such as highlighting the tornado's pressure and rotation rate. The spheres in the low pressure vortex represent the developing tornado, which rise in the updraft and are colored by pressure (see Figs. 5.1 and 5.2). Tilting cones colored based on temperature represent the wind speed and direction at the surface, and show the interaction of warm and cold air around the developing tornado (Wilhelmson et al. 2010).

Donna Cox, leader of NCSA's experimental technologies division, captures quite poetically the team-effort involved in developing this visualization. She pictures her colleagues as "a very hard-working, collaborative renaissance team" (Barker 2004). By saying this, Cox is making explicit that this particular visualization is a complex process, one that involves consultation across researchers and disciplines. Barker further explains that at each stage of the process, researchers are required to consult each other in order to make informed decisions about selecting the most descriptive data, and how best to draw meaningful information from them.

At this point it is interesting to note that a counter-rotating satellite tornado appears of the side of the main tornado (see Fig. 5.2). This is a phenomenon that the experts report as rarely observed in Nature, and it has been infrequently recorded by storm chasers. In fact, the second counter-rotating satellite tornado was not observed in Manchester, nor do the researchers expected its appearance. Nonetheless, the tornado chasers acknowledge the possibility of a satellite tornado appearing given the right assumptions built into the computer simulation. We could conclude that this second tornado *confuses* the researchers in the sense that it is unexpected yet empirically possible. A brief reminder is in order. In Chap. 3, we discussed Mary Morgan's claims that computer simulations *surprise* researchers but do not *confound* them because the behavior of the simulation can be traced back to, and explained in terms of, the underlying model. The reader is now invited to revisit that chapter in light of this example.

Returning to the analysis of visualizations in computer simulations, we see that researchers could create a very complex computer simulation of a tornado in 'real-time,' and visualize it along with its properties and behavior. Through these visualizations, researchers are able to understand the formation, evolution, and demise of a tornado far more efficiently than via any other linguistic form (e.g., mathematical and simulation models, matrices, vectors, or any numerical form of representing results of computer simulations). Researchers are also able to account for the unexpected counter-rotating satellite tornado, predict its evolution, describe its performance,

---

[26]For a full video showing the development of the tornado, see http://avl.ncsa.illinois.edu/wp-content/uploads/2010/09/NCSA_F3_Tornado_H264_864.mov.

Fig. 5.1 Image of the visualization of a F3 type tornado with formation of clouds. Created by the advanced visualization laboratory at NCSA. Courtesy of the National Center for Supercomputing Applications (NCSA) and the board of trustees of the University of Illinois

study its trajectory, and analyze the initial conditions that made possible this satellite tornado in the first place.

Thus understood, this visualization is useful for several epistemic endeavors such as explaining a series of tornado-related questions, measuring their internal values, and predicting the tornado's potential damage. Let us note that the sense given to explanation and prediction here is of being epistemic functions that depend on a non-linguistic base. In this sense, these forms of explanation and prediction depend on a psychological and aesthetic perception of the tornadoes, rather than a linguistic reconstruction of the visualization—if that is possible at all. Another important use

Fig. 5.2 Image of the visualization of a F3 type tornado with the formation of a satellite tornado.
Created by the advanced Visualization Laboratory at NCSA. Courtesy of the National Center for
Supercomputing Applications (NCSA) and the board of trustees of the University of Illinois

of this visualization is for validating the initial input data as well as conferring
reliability to the underlying simulation model. If the tornado's behavior departs too
much from a real tornado—or from whatever the experts anticipate—then researchers
have reasons to believe that the simulation, the input data, or both are incorrect.

The visualization of the tornado—represented statically in Figs. 5.1 and 5.2—
can be summarized as a 3D, dynamic evolution with the possibility to manipulate
the initial and boundary conditions. A characteristic of this kind of visualization,
along with most visualizations in computer simulations, is that they are shown on a
computer screen. Although pointing out something so obvious could look arbitrary, it
comes with specific limitations on the researcher's capacity to manipulate, visualize
and ultimately understand the visualizations. I will return to this point at the end of
the section.

More sophisticated forms of visualizations are found in high-end scientific research facilities. Two forms come to mind, namely, 'virtual reality' (VR) and 'augmented reality' (AR). The first one refers to visualizations where the researcher is able to interact with them by using special gadgets, such as goggles and a 'mouse-wand.' At the High Performance Computing Center Stuttgart (HLRS) at the University of Stuttgart, researchers have recreated an entire replica of Forbach, a town in the Black Forest, along with the Rudolf Fettweis power station. Projecting the visualization on a special room—known as Cave Automatic Virtual Environment (CAVE)[27]— researchers can walk around Forbach, get inside the dam as well as the underground turbines, and inspect the reservoir, all with the help of a special set of lenses. Furthermore, researchers could enter any house in the area and observe how construction and modernization projects affect citizens, wild life, and the environment in general (Gedenk 2017). The fact that the simulation is, so to speak, outside the computer screen, introduces significant advantages in terms of the practice of computer simulations, as well as understanding the results and communicating them.

The second form of visualization is known as 'augmented reality' and, as the name suggest, consists in aggregating simulated information onto the real-world. Interestingly, there are many ways to do this. At the HLRS, a typical case of AR consists of overlying pre-simulated computations onto a tagged real entity. With the help of advanced technology, such as code markers and special cameras, computer vision and object recognition, the amount of insight and understanding of an object in the real-world can be significantly increased.

A simple but illustrative example of AR is the water flow simulated interactively and displayed around a Kaplan turbine (see Fig. 5.4). To this end, researchers need first to pre-simulate the water flow around the turbine using standard equations of flow dynamics—at the HLRS researchers use Fenfloss, a fast Navier-Stokes solver which computes the water flow. The simulated turbine, on the other hand, is a model of the architecture of the real turbine, and thus it needs to be as accurate as possible. Once the pre-simulated data is available, researchers tag with code markers specific places on the real turbine to visualize the simulated flow. With this information, a parametric mesh generator creates the surface and computational mesh of the water turbine. After a few seconds of processing, the results of the simulation are displayed on the real turbine mimicking the real flow.[28]

Let us now ask, what sort of epistemological advantages do VR and AR have to offer?[29] As should be expected, each form of visualization has different epistemic value and thus provides different forms of understanding. A common denominator, however, is that both VR and AR 'naturalize' the simulation in the sense that

---

[27]The CAVE is a three by three by three meter pitch black room with five single-chip DLP projectors with a resolution of 1920 by 1200 pixels, each sending a respective image creating an accurate rendering for the human eye. Four cameras are installed at the corners of the ceiling for tracking the researchers inputs by the glasses and the mouse-wand.

[28]There are a few cases where the simulation is computed in real time during the AR session. The main concern with this kind of technology, however, is that it is too slow and time consuming.

[29]I thank Thomas Obst and Wolfgang Schotte at the HLRS for explaining to me the details of their interesting work.

researchers manipulate the visualization as if they were manipulating a real thing in the world. One could say that the simulation becomes a natural extension of the world, a 'piece' of it in a similar sense ascribed to experimentation. For VR, the naturalization comes in the form of 'walking' inside the simulation, 'looking' below and above, 'touching' objects, 'changing' their location, etc. For AR, the naturalization stems from embedding the simulation into the real-world, and the real-world into the simulation. The simulated flow appears as the natural flow, and the real turbine becomes part of the simulation.

As a consequence, the results of the simulation become more epistemically accessible than a mere visualization on the computer screen, however realistic and sophisticated it might be. AR and VR bring the simulation into the world, and the world into the simulation. They become embedded one into the other, mixing the real and the simulated into one new naturalized form of reality. The success of AR and VR, in sum, is that they demand much less cognitive effort from researchers—as well as policy makers, politicians, and the general public—as they make simulation practice a rather natural scientific and engineering experience.

Let us first focus on VR. As Fig. 5.3 shows, the two researchers are standing by a water surface in a very realistic way. The image also shows the researchers holding a special 'mouse-wand' that helps them navigate through the visualization as well as the menu displayed on the right side of the CAVE. The use of 3-D glasses is also important, as they help orient the visualization in such a way that it appears natural to the human eye.

Fig. 5.3 The Cave Automatic Virtual Environment (CAVE). The CAVE gives researchers a fully immerse, 3-D simulation environment to analyze and discuss their computations. Created by the department of visualization at the high-performance computing center Stuttgart (HLRS), University of Stuttgart. Courtesy of the HLRS, University of Stuttgart

By means of these gadgets, navigation is simplified. The mouse-wand allows the researcher to 'walk' in the simulation, and the 3-D glasses permit researchers to 'look' behind, above, or below different simulated objects in a similar fashion to how they would in real life. As mentioned earlier, the incontestable advantage of using AR is that it brings. As mentioned earlier, the incontestable advantage of using AR is that it brings the simulation into the real-world. But there are also several other advantages of AR meant to facilitate our understanding of a given target system.

In conversation at the HLRS, Thomas Obst and Wolfgang Schotte tell me that AR is very successful in internalizing policy makers, politicians, and the general public into the results of the simulation. Consider the example of the turbine again (Fig. 5.4). It is possible to show the results of the simulations to the relevant stakeholders without the need to be in the CAVE. Instead, the results of a simulation can be visualized in situ with the turbine, a laptop computer and a camera. Portability matters. In fact, AR facilitates the explanation of complex technical issues in a simple and organic manner to researchers that have not been originally involved in the simulation. "Researchers, but also politicians and the general public" says Obst, "can relate and understand the results of the simulation in a much easier way with AR than on a computer screen, or even the CAVE."

Unfortunately, AR has its limitations. A core issue that worries many researchers is that AR depends on a pre-computed stage, that is, the visualized results are not being computed in real time but are rather pre-computed. Because of this, its contribution back to the simulation is limited in several ways. For instance, no real-time errors can be detected in an AR environment. Furthermore, if the design conditions change

Fig. 5.4 Water flow is simulated interactively and displayed on top of a Kaplan turbine. Created by the department of visualization at the high-performance computing center Stuttgart (HLRS), University of Stuttgart. Courtesy of Stellba Hydro GmbH & Co KG

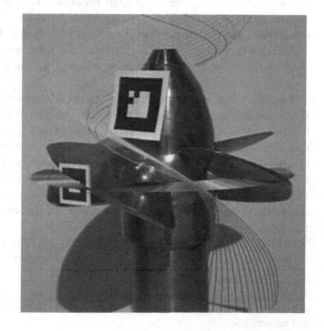

(e.g., elements have been added or subtracted from the original simulation or the material setup), then the AR environment could be rendered utterly useless.[30]

An important benefit of VR and AR over visualizations on the computer screen is that they are neither forcing the researchers nor the public—politician, policy makers, etc.—into one fixed perspective over another. This point has to do with the 'naturalization' of a visualization brought up by VR and AR mentioned earlier. Researchers and the public are able to manipulate the simulation, and thus focus their attention on what is of most importance to them. Contrast this with a visualization on a computer screen. The use of a mouse—or perhaps a touch screen—limit the points of attention, sets the order of importance, and contrive to a given perspective. While using visualizations on a screen to communicate results to the public, researchers have a pre-chosen perspective of what to show (e.g., by selecting the angle, navigating through the options menu and so on). Instead, when the communication of results is via VR or AR, the public is able to interact with the visualization in a different way, and thus understand it on its own terms, without any pre-chosen perspective.

In a straightforward sense, with VR and AR researchers and the public do not need to know much about the simulation models nor their assumptions in order to understand the results of computer simulations. Whereas this is also true of visualizations on the computer screen, VR and AR naturalize the visualization experience.

5.3 Concluding Remarks

Throughout this book, I have argued that computer simulations play a central role in contemporary science and engineering research. Their importance lies in the epistemological power they provide as methods of research. In the previous chapter, I argued that computer simulations provide knowledge about a target system. In this chapter, I show how understanding is achieved by using computer simulations in scientific and engineering research.

I then divided the chapter into two forms of understanding, namely, *linguistic forms* and *non-linguistic forms* of understanding. The distinction is intended to show that there are forms of understanding that depend on a symbolic form (e.g., formulae, specifications, and the like) and other forms of understanding that depend on non-symbolic forms (e.g., visualizations, sounds, interactions). Cases that qualify as the former are the explanatory force of computer simulations, their predictive use, and the possibility to generate significant findings about real-world phenomena. Each one of these depends, in one way or another, on the simulation model as a set of formulae.

As for the non-linguistic forms of understanding, I addressed the case of visualizations in computer simulations. It is well known among the scientific and engineering

---

[30]These issues could be overcome by computing and visualizing results in the AR environment in real time. Unfortunately, this kind of technology has high costs in terms of computing process, time and memory storage.

communities that visualization is an important vehicle for understanding the results of computer simulations. In the section dedicated to this topic, I discuss standard visualizations on computer screen, and the less common yet central for scientific and engineering purposes, virtual reality and augmented reality.

Chapter 6
Technological Paradigms

In previous chapters, we have discussed how philosophers, scientists, and engineers alike construct the idea that computer simulations offer a 'new epistemology' for scientific practice. By this they meant that computer simulations introduce new—and perhaps unprecedented—forms of knowing and understanding the surrounding world, forms that were not available before. Whereas scientists and engineers emphasize the *scientific* novelty of computer simulations, philosophers try to appraise computer simulations for their *philosophical* virtues. The truth of the former claim is out of question, the latter, however, is more controversial.

Philosophers have argued for the novelty of computer simulations on several occasions. Peter Galison, for instance, defends the idea that "physicists and engineers soon elevated the Monte Carlo above the lowly status of a mere numerical calculation scheme, [as] it came to constitute an alternative reality—in some cases a preferred one—on which 'experimentation' could be conducted. Proven on what at the time was the most complex problem ever undertaken in the history of science—the design of the first hydrogen bomb—the Monte Carlo ushered physics into a place paradoxically dislocated from the traditional reality that borrowed from both experimental and theoretical domains, bound these borrowings together, and used the resulting bricolage to create a marginalized netherland that was at once nowhere and everywhere on the usual methodological map" (Galison 1996, 119–120).

Naturally, Galison is not alone in his claims. Many others have also joined him arguing in favor of the epistemological, methodological, and semantic value of computer simulations. Fritz Rohrlich, for instance, is one of the first philosophers to locate computer simulations in the methodological map, lying somewhere between theory and experiment. He says "computer simulation provides [...] a qualitatively new and different methodology for the natural sciences, and [...] this methodology lies somewhere intermediate between traditional theoretical science and its empirical methods of experimentation and observation. In many cases it involves a *new syntax* which gradually replaces the old, and it involves *theoretical model experimentation* in a qualitatively new and interesting way. Scientific activity has thus reached a new

© Springer Nature Switzerland AG 2018

J. M. Durán, *Computer Simulations in Science and Engineering*,
The Frontiers Collection, https://doi.org/10.1007/978-3-319-90882-3_6

milestone somewhat comparable to the milestones that started the empirical approach (Galileo) and the deterministic mathematical approach to dynamics (the old syntax of Newton and Laplace). Computer simulation is consequently of considerable philosophical interest" (Rohrlich 1990, 507, italics original). Authors like Eric Winsberg have also claimed that"[c]omputer simulations have a distinct epistemology [...] In other words, the techniques that simulationists use to attempt to justify simulation are unlike anything that usually passes for epistemology in the philosophy of science literature" (Winsberg 2001, 447). He is also known for saying that "computer simulations are not simply number crunching techniques. They involve a complex chain of inferences that serve to transform theoretical structures into specific concrete knowledge of physical systems [...] this process of transformation [...] has its own unique epistemology. It is an epistemology that is unfamiliar to most philosophy of science" (Winsberg 1999, 275). Similarly, Paul Humphreys considers computer simulations as being essentially different from the way we understand and evaluate traditional theories and models (Humphreys 2004, 54). The list of authors goes on.

Our treatment of epistemic functions in Chap. 5 reveals much of the form that epistemological studies take for computer simulations. As presented and discussed, several epistemic functions are performed well by computer simulations, rendering understanding of a given target system. Explanation, prediction, exploration, and visualizations are just a handful of these functions with special epistemological input for scientific and engineering practice. However, many more epistemic functions could also be mentioned. Think of new forms of observation and measuring the world through computer simulations, modes for evaluating evidence, and ways to confirm/refute scientific hypothesis, among others. Molecular dynamic simulations in chemistry, for instance, now have the potential to give valuable insights into experimental results. And simulations of quantum states allow scientists to select potential atom types for specific targets, even before sitting at the laboratory bench (e.g., Atomistix Toolkit). All these uses and functions of computer simulations are, nowadays, quite common and to a reasonable extent, critical to the progress of modern science and engineering.

We also saw that, in an increasing number of occasions, simulations antecede experiments. Molecular dynamic simulations again furnish useful examples. These simulations are able to provide simple chemical pictures of ionized intermediates and reaction mechanisms essential for different origin-of-life scenarios. The very same simulations help identify atomic-scale properties that determine observed macroscopic kinetics. With the presence of these simulations, experiments become more tractable and accurate, as the simulation helps to narrow down the number of materials used, reactive conditions, and density configurations. In addition, these simulations facilitate the manipulation of time and length scale constraints that would otherwise limit molecular experimentation. In this respect, computer simulations complement scientific and engineering practice by making it possible.

In this context, some philosophically well-versed scientists have argued that computer simulations constitute a new paradigm of scientific and engineering research. The first paradigm being theory—and modeling—whereas the second paradigm is laboratory and field experimentation. In this chapter we will discuss what it would entail to call computer simulations a third paradigm of research. Similarly, Big Data has been called the 'fourth paradigm' of scientific research, making it impossible to overlook the symmetry with computer simulations. Because of this, in this chapter I will draw some similarities and differences between one and the other, with the hope of understanding their scope and limits.

6.1 The New Paradigms

The denomination of computer simulations as a third paradigm of research has its origin within the advocates of Big Data studies. In fact, partisans of computer simulations have never referred to them in this way, despite strong advances in favor of their scientific and philosophical novelty.

Conceiving computer simulations as a paradigm of research—regardless of their ordinal position—adds some pressure to the expectations that researchers and the general public have on them. Physics is conceived as a paradigm for all natural occurring phenomena because it provides insight into such phenomena. The electromagnetic theory, for instance, is capable of explaining all electro and magnetic phenomena, and general relativity generalizes special relativity and Newton's law of universal gravitation providing a unified description of gravity as a geometric property of space and time. To call computer simulations and Big Data a paradigm, one might think, must have a similar epistemic value as physics in terms of the insight they provide as well as their role as an epistemic authority. It is therefore important to briefly discuss the extent of such a (meritorious) title.

Jim Gray suggested calling computer simulations and Big Data new paradigms of research[1] in a talk given to the National Research Council and to the Computer Science and Telecommunications Board in Mountain View, CA on January 2007. During his talk, Gray connected computer simulations and Big Data with a chief concern existing in contemporary research, that is, that scientific and engineering practice are being affected by a 'data deluge.' The metaphor was intended to call attention to genuine concerns that many researchers have regarding the large amount of data stored, rendered, gathered, and manipulated by science and engineering practitioners. To give a simple example, the Australian Square Kilometer Array Pathfinder (ASKAP) consists of 36 antennas, each of 12 m in diameter, spread out over 4,000 square meters, and working together as a single instrument rendering as much as

---

[1]For the most part, I am going to ignore the ordinal position of the paradigm. In this respect, I will leave unanswered the question of whether there is some presupposed hierarchy among the paradigms. As we saw in Chap. 3, the positivist would take that experimentation remains a secondary, confirmatory stance of theory, where theory is more important. After the flaws of the positivist were exposed, a new experimentalist wave invaded the literature showing the vast universe of experimentation and their philosophical importance. Should advocates on computer simulations and Big-Data claim that there is a new, improved way of doing science and engineering, and that theory and experimentation are reserved for only a minor role, they would be walking down the same dangerous road as the positivist.

700 TB/s of data.[2] Thus understood, the data deluge stems from an excess in the production and collection of data that no single team of researchers can process, select, and understand without further aid by a computational system. This is the principal reason why Big Data requires special algorithms that help to sort out what is important data and what is not (e.g., noise, redundancies, incomplete data, etc.).

In this context, it is important to elucidate the notion of 'paradigm' as used by Grey and supporters because, so far, there are not available definitions. In particular, this is a term that can not be taken lightly, especially if it has implications in the status—cultural, epistemological, social—of a discipline and the way the public will accept it. In philosophy, 'paradigm' is a theoretical term that comes with specific assumptions. Clarifying the meaning of a 'paradigm' in the context of computer simulations and Big Data is, then, our first task.

Before we begin, one caveat must be acknowledged. Earlier, we discussed hybrid scenarios where computer simulations integrate with laboratory experimentation. In Big Data studies we face a similar situation. The Large Hadron Collider (LHC) is a good example of such an integration, as it exquisitely combines cutting edge science and technology with large amounts of data including, of course, experimentation, theory, interdisciplinary work, and computer simulations. Many simulations at the LHC are meant to optimize the computing resources needed to model the complexity of detectors and sensors, as well as the physics (Rimoldi 2011). Others, like state-of-the-art Monte Carlo simulations, calculate the Standard Model Higgs boson signal and any relevant background processes. The use of these simulations is to optimize selections of events, to evaluate their acceptance, and assess systematic uncertainties (Chatrchyan et al. 2014). These simulations are meant to produce large amounts of data that eventually need to be carefully curated, selected, and classified. Thus understood, computer simulations and Big Data are deeply interwoven in the process of rendering data, classifying it and using it, among other activities. In this book, I have intentionally avoided discussing hybrid scenarios such as the LHC suggests. This decision is based on a very simple reason. To me, before we are able to grasp computer simulations as hybrid systems, it is important to first understand them individually. Whereas hybrid scenarios provide a richer view, they also obscure important aspects of the epistemological and methodological analysis. In this respect, when discussing the third and fourth paradigm of research, I will be addressing computer simulations and Big-Data in a non-hybrid scenario.

Thomas Kuhn is the first philosopher to analyze in depth the idea of a 'paradigm' in scientific research. When discussing how paradigms function in science, Kuhn notes that "one of the things a scientific community acquires with a paradigm is a criterion for choosing problems that, while the paradigm is taken for granted, can be assumed to have a solution. To a great extent these are the only problems that the community will admit as scientific or encourage its members to undertake" (Kuhn 1962, 37). Calling computer simulations and Big-Data a paradigm of scientific research has

---

[2]Although there are several projects in science and engineering relying on Big Data, its presence is much stronger in areas such as social networking studies, economy, and big government.

specific implications in the way researchers carry out their practice, the problems worth solving, and the right methods to pursue such solutions.

What is, then, a 'paradigm'? According to Kuhn, any mature science (e.g., physics, chemistry, astronomy, etc.) experiences alternating phases of normal science (e.g., Newtonian mechanics) and revolutions (e.g., Einsteinian relativism). During periods of normal science, a constellation of commitments are fixed, including theories, instruments, values, and assumptions. These conform a 'paradigm,' that is, the consensus on what constitutes exemplary instances of good scientific research. The function of a paradigm, then, is to supply puzzles for scientists to solve and to provide the tools for their solution (Bird 2013). As examples, Kuhn cites the chemical balance found in the *Traitè élémentaire de chimie* by Antoine Lavoisier, the mathematization of the electromagnetic field by James Clerk Maxwell, and the invention of calculus in *Principia Mathematica* by Isaac Newton (Kuhn 1962, 23). Each one of these books contain not only the key theories, laws, and principles of Nature, but also—and this is what makes them paradigms—guides on how to apply those theories for the solution of important problems (Bird 2013). Furthermore, they also provide new experimental and mathematical techniques for the solution of such problems. Examples of this sort have been already mentioned: the chemical balance for the *Traitè élémentaire de chimie* and the calculus for the *Principia Mathematica*.

A crisis in science arises when the scientists' confidence in a paradigm is lost due to its inability of, or failure to, solve a particularly worrying puzzle. These are the 'anomalies' that emerge in times of normal science. Such crises are typically followed by a *scientific revolution*, having one existing paradigm superseded by a rival. During a scientific revolution, the disciplinary matrix (i.e., the constellation of shared commitments) undergoes revision, sometimes even shaking to the core the corpus of beliefs and world view. Such revolutions typically emerge from the need of finding new solutions to anomalies and disturbing new phenomena that were coexisting within theories in periods of normal science. The classical example is the precession of the perihelion of Mercury that worked as a confirmatory stance for general relativity over Newtonian mechanics.[3] Revolutions, however, do not necessarily affect the scientific progress, mainly because the new paradigm must retain at least some core aspects of its predecessor, especially the power to solve quantitative problems (Kuhn 1962, 160ff.). It is possible, however, that the new paradigm may lose some qualitative and explanatory power (Kuhn 1970, 20). In any case, we can say that in periods of revolution, there is an overall increase in the puzzle-solving power, the number and significance of the puzzles, and the anomalies solved by the revised paradigm (Bird 2013).

A paradigm, then, informs scientists about the scope and limits of their scientific domain, at the time that warrants that all legitimate problems can be solved within its own terms. Thus understood, it seems that neither computer simulations nor Big Data fit into this description. For starters, they are *methods* that implement theories and

---

[3]The precession of the perihelion of Mercury was explained by general relativity around 1925—with successive and more accurate measurements starting in 1959—although it was an 'anomalous' phenomenon already known back in 1919.

models, but not theories in themselves, and therefore they are not suited to promote a scientific crisis. Could they bestow a theory that questions our basic understanding of, say, biology? Probably, but in this case they would not have a different status than any of the experiments used for debunking theories about spontaneous generation of complex life from inanimate matter.[4] One could of course speculate that computer simulations and Big Data might become, in and by themselves, theories of some sort—or means for a theory. It is true that some philosophers have declared the 'end of theory' brought about by Big Data, but there is little evidence that research practice is actually heading in that direction. Moreover, changes of paradigm come with the new paradigm withholding the same explanatory and predictive force as the old, with the addition of taking care of the anomalies that lead to the crisis in the first place. Big Data, as we will see in this chapter, not only has little preoccupation with 'accumulating' from previous paradigms, but it openly rejects many of its triumphs. Most evidently is the rejection of the need for *explanations* of phenomena of any kind. As many advocates of Big Data admit, it is not possible to explain *why* real-world phenomena happen, but only to show *that* they happen.

What, then, had Gray in mind when he called computer simulations and Big Data the third and fourth paradigm, respectively? Let us begin by noticing his division of scientific paradigms into four historical moments, namely,

1. Thousand of years ago, science was *empirical* describing natural phenomena;
2. Last few hundred years *theoretical* branch using models, generalizations;
3. Last few years a *computational* branch simulating complex phenomena;
4. Today: *data exploration* eScience unify theory, experiment, and simulation
   - data captured by instruments or generated by simulator,
   - processed by hardware,
   - information/knowledge stored in computer,
   - Scientists analyses database/files using data management and statistics (Gray 2009, xviii).

Under this interpretation, a 'paradigm' is not so much a technical term in the sense given by Kuhn, as it is the set of coherent research practices—including methods, assumptions, and terminology—that a community of scientists and engineers share among themselves. Such research practices do not require a scientific revolution, nor promote one. In fact, since computer simulations and Big Data make use of current scientific and engineering standards, theories, and the like, they seem to be already inserted into a paradigm. The hallmark of computer simulations and Big Data, however, is the mechanization and automatization of data by computers, clearly lacking in the first two paradigms. This means that the methods used and offered in the third and fourth paradigm are significantly different from experimenting with phenomena and theorizing about the world. I will refer to computer simulations and

---

[4]Louis Pasteur showed that the apparent spontaneous generation of microorganisms was actually due to unfiltered air allowing bacterial growth.

Big Data as 'technological paradigms' in an attempt to put some distance from the philosophical interpretation of 'paradigm' presented by Kuhn. Let us now see if we can make sense of these new technological paradigms.

6.2 Big Data: How to Do Science with Large Amounts of Data?

As we understand the term today, Big Data[5] refers to a large data set whose size goes well beyond the ability to capture, manage, and process data by a cognitive agent within a reasonable elapsed time. Unfortunately, there is not a proper conceptualization of Big Data whose notion is typically captured by listing some of its ascribed features, such as large, diverse, complex, longitudinal, etc. The core problem with having a list of features, characteristics, and attributes is that they do not necessarily illuminate the concept. To say that it has four legs, it is furry, and barks does not illuminate the concept of a 'dog.' In particular, such lists tend to obscure rather than shed light on the meaning of Big Data as we have not advanced one single step by defining 'big' in terms of 'large,' 'diverse,' and the like. A better way to characterize Big Data is needed.

In 2001, Douglas Laney proposed a definition of Big Data based on what he called the 'three Vs': *volume, velocity*, and *variety* (Laney 2001). Unfortunately, this definition did not work too well. Not one of the 'V' provides any real insight into the notion, the practice, and the uses of Big Data. Later in 2012, the definition was refined by Mark Bayer and Laney himself as "high volume, high velocity, and/or high variety information assets that require new forms of processing to enable enhanced decision making, insight discovery and process optimization" (Beyer and Laney 2012). Better suited, but still unfitting. Rob Kitchin further extended the list of characteristics and functions that constitutes Big Data, including exhaustive in scope, fine-grained in resolution, and uniquely indexical in identification, relational, flexible, etc. (Kitchin 2014). Unfortunately, neither of these definitions shed any more light on the notion of Big Data than a plain list. Notions such as 'volume' and 'variety' do not advance our understanding of 'big' and 'large.' A bag of candy might be varied and in large volume, and yet it says nothing about the candy itself.

---

[5] Wolfgang Pietsch has suggested a distinction between Big Data and *data-intensive science*. While the former is defined with respect to the amount of data and the technical challenges which it poses, data-intensive science refers to "to the techniques with which large amounts of data are being processed. One should further distinguish methods of data acquisition, data storage, and data analysis" (Pietsch 2015). This is a useful distinction for analytical purposes, as it allows philosophers to draw conclusions about data regardless of the methods involved; in other words, distinguishing Big Data as a product from a discipline. To us, however, this distinction is otiose because we are interested in studying the techniques of data acquisition against a backdrop of technical components (e.g., speed, memory, etc.). A similar remark applies to Jim Gray's notion of *eScience*, understood as "where IT meets scientists" (Hay et al. 2009, xviii). In the following, although I make use of the notion of Big Data, data intensive science, and eScience indistinctively, readers must keep in mind that they are different fields.

But there is more. Terms like 'big,' 'large,' 'abundant,' and others can be constituents of a *relational predicate*, that is, as part of a comparison relationship: one-hundred meters is a large street block; and a one-thousand page book is a big book. But none of these predicates are absolute. A one-hundred meters street block could be large for a German living in an old, medieval-style city. But it is of normal size for an Argentinean citizen where most street blocks are about one-hundred meters. In other words, what is big to you might not be big to me (Floridi 2012). Of course this is not a mere cultural or societal factor. Technological change is especially sensible to relational predicates. What could have been considered "Big Data" in the middle 1950s is, by all accounts, insignificant today. Compare for instance the amount of data for the 1952 US presidential election, which made use of the UNIVAC with a storage capacity of about 1.5 MB per tape, with the 700 TB/s produced by the Australian Square Kilometer Array Pathfinder (ASKAP). Under the umbrella of both being 'big,' the US election and the ASKAP can be methodologically and epistemically treated as par, when obviously they should not be. The lesson to draw here is that solely listing properties and attributes about the data fails to provide insight into the concept, practices and uses of Big Data.

Before continuing, let us take one step back and ask: why is it so important to have a definition of 'Big Data'? One possible answer is that lacking a conceptual and theoretical specification of this notion might have serious implications in the regulation of technological practices, at individual, institutional, and governmental dependency levels. Consider the following description for funding projects by the National Science Foundation (NSF): "The phrase 'Big Data' in this solicitation refers to large, diverse, complex, longitudinal, and/or distributed data sets generated from instruments, sensors, Internet transactions, email, video, click streams, and/or all other digital sources available today and in the future" (NSF 2012). If anything, this description is unclear and vague, and it does not help to specify what constitutes Big Data, its aims and limits, and its purpose, all being fundamental information for applicants and executors of the grant. Moreover, the overuse of synonyms does not deliver by itself clarification on what 'Big Data' refers to, nor what it is its general scope. Compared to the document that replaced the solicitation two years later, one reads again with great ambiguity: "The phrase 'Big Data' refers to data that challenge existing methods due to size, complexity, or rate of availability" (NSF 2014). It is interesting to note that in the two following solicitations—2015 and 2016—the NSF eliminated all references to definitions of Big Data. At best, one can find statements indicating "[w]hile notions of *volume*, *velocity*, and *variety* are commonly ascribed to Big Data problems, other key issues include data quality and provenance" (NSF 2016) in clear reference to Laney's 2001 definition. These solicitations by the NSF shows how difficult it is to define, or even clearly characterize, the notion of Big Data.

To the best of my knowledge, there is no unproblematic, all-inclusive definition of Big Data. This fact, however, does not seem to preoccupy the majority of the specialized literature. Some authors argue that the novelty of Big Data lies in the sheer quantity of data involved, taking this as an intrinsically intuitive and sufficiently valid characterization. In a similar sense, it has been claimed that Big Data refers to datasets whose size goes beyond the ability of a typical commercial database software to capture, store, manage, and analyze the data. Instead, specific storage devices and software need to be used. At its core, the amount of data dealt with range from a few dozen Petabytes to thousands of Zettabytes, and even higher. As naturally as they come, these numbers change as new computers enter the technological scene by allowing more storage and computational processing speed, as new algorithms are developed for sorting and categorizing the data, and as new institutions and private corporations invest money in facilities and personnel for developing Big Data even further. To put these numbers into real values and applications, it is estimated that about 2.5 Exabytes of data is generated every single day—that is 2.5×10^{18} Bytes. Mark Liberman, a linguistic professor at the University of Pennsylvania, has estimated that the storage required for keeping all human speech digitized at a 16 kHz, 16-bit audio, is about 42 Zettabytes—or 42×10^{21} Bytes—(Liberman 2003). Furthermore, since the global production of computer-based data is growing at unprecedented speed, predictions estimate that at least 44 Zettabytes—that is 44×10^{21} Bytes will be produced by 2020 (IDC 2014).[6]

Accepting the fact that there is no clear definition of what constitutes Big Data should not suggest that we cannot pick out and highlight specific characteristics of it. Obviously, the sheer quantity of data is indeed a distinctive characteristic that allow us to discuss in more detail the methodology and epistemology of Big Data. Furthermore, Sabina Leonelli in a remarkable article has correctly pointed out that the novelty of Big Data lies in the prominence and status acquired by data in the sciences, as well as in the methods, infrastructure, technologies, and skills developed to handle such data (Leonelli 2014). Leonelli is right to point out that Big Data raises data to the status of 'scientific commodity,' in the sense that equates other units of analysis such as models and theories in their scientific relevance. She is also right to say that Big Data introduce methodologies, infrastructure, technology and skills that we did not have before. Big Data, then, is about sheer quantity of data, new methods and infrastructure, as well as the development of new technology. But above all, Big Data is an entirely new and different way to practice science and understand the surrounding world, very much like computer simulations. I believe this is the sense that Gray had in mind when he called computer simulations and Big Data new paradigms of scientific research.

So far, our efforts have been focused on trying to conceptualize Big Data. Let us now have a look at its actual practice in scientific and engineering domains. A short reconstruction of Big Data would include some practices at the gathering stage, some at the curating stage, and some at the use of the data. Let us begin with the

---

[6]There are studies that link the growth of the system memory with the amount of data stored. A report to the Department of Energy of the US shows that, on average, every 1 Terabyte of main memory results in about 35 Terabyte of new data stored to the archive each year—more than 35 Terabyte per 1 Terabyte actually get stored, but 20–50% actually gets deleted on average over the year (Hick et al. 2010).

first stage, when data is collected by different computational means.[7] The example of the ASKAP above provides a good case in point. In there, data is obtained by one single array of radio telescopes and formatted in ways that make them compatible with different datasets and standards. Obtaining data from one single constellation of instruments, however, is not a typical case for Big Data. More usual is to find several different and, most likely, incompatible sources contributing to enlarge the databases. One could think of combining data from the ASKAP, which is a radio telescope for exploring the origins of galaxies, with data obtained from the University of Tokyo Atacama Observatory (TAO), an optical-infrared telescope for understanding the Galactic center (Yoshii et al. 2009).[8] Gathering data requires making sure that they are formatted in ways that make data compatible with actual datasets, as well as making sure that the metadata is standardized in proper ways. Formatting data, as well as standardizing the metadata, is a time-consuming and expensive business, although necessary to ensure that the availability of the data and that they can eventually be analyzed as a single body of information.

Here is where the first set of problems for Big Data emerges. Having a constellation of sources present significantly enhances the problems for sharing data, their formatting and standardization, curation, and eventual use. Since at this initial stage core issues are sharing, formatting, and standardizing data, let us focus on those first. Leonelli reports that it is frequent that researchers, who wish to share their data to a database, need to make sure that the data and metadata they are submitting are compatible with existing standards. This means finding the time in their already busy agenda for acquiring updated knowledge on what the standards are and how they can be implemented (Leonelli 2014, 4). If things were not difficult

---

[7]Our treatment of Big Data will be focused on scientific uses. In this respect, we need to keep in mind that, although the nature of the data is always *computational*, it is also of an empirical origin. Allow me here to make a quick digression and clarify what I mean by computational data of empirical origin. In laboratory experimentation, the data gathered could come directly from manipulating the experiment and thus by means of reporting changes, measurements, reactions, etc., as well as by means of using laboratory instruments. An example of the former is using a Petri Dish to observe the behavior of bacteria and plant germination. An example of the latter is the bubble chamber that detects electrically charged particles moving through a superheated liquid hydrogen. Big Data in scientific research obtains much of its data from similar sources. As suggested, the ASKAP obtains large quantities of data from scanning the skies, and in this respect data has an empirical origin. Even the data collected from social networks largely used for sociologists and psychologists could be considered to have an empirical origin. To contrast these ways of gathering data, let us take the case of computer simulations. With the latter methods, data is *produced* by the simulation rather than collected. This is not a whimsical distinction, since the characteristics, epistemological assessment, and quality of the data varies significantly from method to method. Philosophers have been interested in the nature of data and on what make them different (for instance Barberousse and Marion 2013; Humphreys 2013a, b).

[8]An interesting case stems from biomedical Big Data where data is gathered from an incredibly varied and complex variety of sources. As Charles Safran et al. indicate, these sources include "laboratory auto-analyzers, pharmacy systems, and clinical imaging systems [...] augmented by data from systems supporting health administrative functions such as patient demographics, insurance coverage, financial data, etc. ... clinical narrative information, captured electronically as structured data or transcribed 'free text' [...] electronic health records" (Safran et al. 2006, 2).

enough, neither academia nor public and private institutions always appreciate their researchers spending qualitative time on such endeavors. As a result, unacknowledged efforts provide researchers very few incentives to share, format, and standardize their data, with the subsequent loss of professional collaboration and research efforts.

However labor-intensive and costly gathering data might be, if Big Data is to be a reality then the development of databases, institutions, legislations, and long-term research funding and trained personnel is necessary. This point brings us to the next step in the process of Big Data, involving many of these actors in the curation of the data (Buneman et al. 2008). Typically, curating consists in making specific decisions regarding the data, ranging from what data to collect to their final care and documentation, including activities such as structuring the data, providing assessment of the quality of data, and offering guidelines for researchers interested in large datasets. Curating data is a hard and often a thankless task, since curators must make sure that the data is readily available for the researchers. It is also highly sensitive, for curating presupposes selection, and thus an increase in the idiosyncrasy of the data. Finally, what makes the curation process into a complex and contentious activity is that it partly determines the ways in which data will be used in the future. Despite its importance, and just like gathering the data, curation is not currently an activity rewarded within the academic and institutional system, and therefore not very appealing to researchers.

As suggested by the previous discussion, the data that is made available through databases is not 'innocent' data. It has been selectively gathered, curated, and made available to researchers with specific objectives in mind. This is of course not to say that they are untrustworthy data, but rather that they are necessarily biased, loaded with values—epistemic, cognitive, social, and moral—that need to be displayed in the open before their use. Unfortunately, this last request is rather difficult to comply with. Researchers, just like any other individual, participate in a society, a culture, a set of pre-defined values that they project—sometimes unknowingly—into their practice. Interestingly, it has been argued that Big Data is countering the risk of bias in data collection, curation, and interpretation. This is because having access to large datasets makes it more likely that bias and error will be automatically eliminated by a 'natural' tendency of data to cluster together—this is what sociologists and philosophers call *triangulation*. Therefore, the more data that is collected, the easier it becomes to cross-check them with each other and eliminate the data that looks like outliers (see Leonelli (2014), Denzin (2006), and Wylie (2002)).

When the question on Big Data comes down to the uses of such data, answers scatter in all directions depending on the discipline: medicine (Costa 2014), sustainability studies (Mahajan 2012), biology (Callebaut 2012) and (Leonelli 2014, genomics (Choudhury et al. 2014), astronomy (Edwards and Gaber 2014), just to cite but a few uses of Big Data in scientific contexts (see also Critchlow and Dam 2013). In the next section, I discuss one such use in virology. Above and beyond the many applications of Big Data, the question about its use is predominantly a question about ethics. Questions about the use of Big Data in medicine resort to questions about the patient's consent and anonymisation, as well as the privacy and protection of data (Mittelstadt and Floridi 2016a). Likewise, questions about ownership, intellectual property, and how to distinguish academic from commercial Big Data are fundamental questions of an ethical flavor. The answers to these questions carry the power to harm individuals or to contribute to new discoveries. In this book, I will not be discussing any of these issues. The question about uses of Big Data is, for me, a question about their epistemology. As for issues about ethics, I concentrate efforts in the much less developed field of ethics of computer simulations (see Chap. 7). For further readings on ethics in Big Data, however, I recommend (Mittelstadt and Floridi 2016b; Bunnik et al. 2016; Collmann and Matei 2016).[9]

6.2.1   An Example of Big Data

In a recent acclaimed book, Viktor Mayer-Schönberger and Kenneth Cukier Mayer-Schönberger and Cukier (2013) provide a beautiful example that paints a full-length portrait of Big Data and its potential use for scientific purposes. The story begins in 2009 with the world outbreak of a novel strain of influenza A, the H1N1/09. In situations like this, timeliness and accuracy of information are critical. Delays and imprecisions not only cost lives, but in this particular case threaten the outbreak of a pandemic. Public health authorities in the U.S. and Europe were too slow, taking days and even weeks to get the ins and outs of the situation. Of course, several factors were at play that contributed to the aggravation of the delays. On the one hand, people usually go untreated for several days before going to the doctor's office. Health authorities are then powerless as their records and data depend on collecting public information (i.e., from hospitals, clinics, and private physicians), none of which involves intruding on the privacy of the home. The Centers for Disease Control and Prevention (CDC) in the U.S., and the European Influenza Surveillance Scheme (EISS) relied on the virological and clinical data retrieved from these channels, and therefore the information they held was mostly incomplete for an accurate assessment of the situation. On the other hand, relaying the information from these sources back to the central health authorities could take an awfully long time. The CDC published national and regional data on a weekly basis, typically with a 12-week reporting lag. Retrieving and relaying information was a painful, inaccurate, and very slow process. As a result, information of where there had been an outbreak, as well as an accurate map of potential spreading were, most of the time, either missing or far too unreliable.

In February of the same year, and just a few weeks before the virus hit newspapers headlines, a group of engineers at Google published an article in Nature describing a new way to obtain reliable information that could potentially predict the next outbreak of the virus, as well as to provide an accurate map of the spreading. It was

---

[9]Let us note that the ethical issues raised in each one of these books are not necessarily limited to scientific and engineering uses of Big Data, but they also extend to business, society, and studies on government and law.

known as the *Google Flu Trends* project which advocates a simple but quite novel idea (Ginsberg et al. 2009). According to the Google engineers, the current level of weekly influenza activity in the U.S. could be estimated by measuring the relative frequency of specific Internet queries which, according to the authors, was highly correlated with the percentage of physician visits in which a patient presented with influenza-like symptoms. In other words, they put to use millions of Gigabytes of Internet queries and compared them with CDC data. All of this could be done within the amazing time of one day. Things were speeding up, and if it continued that way, health authorities would be able to address outbreaks within a reasonable time frame.

According to these Google engineers, "Google web search queries can be used to estimate ILI [influenza-like illness] percentages accurately in each of the nine public health regions of the United States" (1012). The message was strong and appealing, and of course everyone took note. Authors like Mayer-Schönberger and Cukier adventured that "by the time the next pandemic comes around, the world will have a better tool at its disposal to predict and thus prevent its spread" (Mayer-Schönberger and Cukier 2013, 10). Expectations are high, and so are the interests for the success of Big Data in scientific disciplines.

Unfortunately, Big Data is all but exact, and the enthusiasm and high hopes raised by these proponents need to be measured. For starters, it is far from clear that Big Data could be used for "the next pandemic," unless some characteristics are shared, such as the kind of pandemic and the ways to retrieve information. Big Data is geographically, economically, and socially dependent. Cholera is currently classified as a pandemic, although it is very rare in developed and industrialized countries, and the way it breaks out and spreads is rather different from an ILI disease. Furthermore, areas with an ongoing risk of disease, such as some places in Africa and south-east Asia, do not have the proper technological infrastructure for the citizens to access the Internet, let alone to query using Google. Retrieving information from these places would most likely not pass minimum requirements of correlation.

One could of course focus only on successful cases of Big Data. The assumption under scrutiny in such cases is that Big Data is a substitute for, rather than a supplement to, traditional data collection and analysis. This is known in the community as 'Big Data hubris,'[10] and aims at highlighting that a history of (relative) success is by no means the history of a method. Even Google's engineers are careful to warn us that the system was not designed to be a replacement for laboratory diagnoses and medical surveillance. Moreover, they admit that the search queries in the model used by Google Flu Trends are not uniquely submitted by users experiencing influenza-like symptoms, and therefore the observed correlations are susceptible to false alerts cause by a sudden increase in ILI-related queries. "An unusual event, such as a drug recall for a popular cold or flu remedy, could cause such a false alert" (Ginsberg et al. 2009, 1014) says Jeremy Ginsberg the lead scientist that published the findings.

---

[10]The term 'hubris' is usually found in the great Greek tragedies describing the hero's personality quality as being of extreme and foolish pride, or holding a dangerous overconfidence often in combination with arrogance. The hero behavior is to defy the established norms by challenging the gods, with the result of the hero's own downfall.

Although Google Flu Trends is just one example of the use of Big Data, and admittedly not the most successful one, it still nicely portrays the advantages of Big Data as well as its intrinsic limitations: more data does not mean better information, it just means more data. Moreover, not all data is 'good quality' data. This is the reason why Ginsberg et al. make clear that using search engine query data is not designed to replace nor supplant the need for laboratory based diagnoses and medical surveillance. Reliable medical data is still obtained by traditional surveillance mechanisms, such as simply going to the doctor. One expected phenomenon related to the model used by Google Flu Trends is that, in a situation of panic and concern, healthy individuals will cause a surge of ILI-related queries boosting the estimates of the actual infected people and potential spread scenarios. This is because queries are not limited to users who are actually experiencing influenza-like symptoms, but rather any individual that posts the right query. The correlations observed, therefore, are not meaningful across all populations. In other words, the system is susceptible to a number of false alerts, not all able to be tracked back and eliminated, rendering the whole program of anticipating a pandemic *via* Big Data still unreliable. The hope is that with the proper corrections and uses of Google Flu Trends, it will provide good insight into the pattern of the spreading of ILI-related diseases, as well as become a useful tool for public health officials to give them an early, more informed response in case of an actual pandemic. To this end, Google Flu Trends has now been refocused on making the data available to institutions that specialize in infectious disease research to use the data to build their own models.

Let us finally note that Google Flu Trends thrives from different sources of data than the ASKAP example. In the former case, the sources are search query data, while in the latter the source is astronomical information provided by radio telescopes. In this sense, the two cases hold many differences, such as the susceptibility of false alerts in the case of Google Flu Trends absent in ASKAP. Furthermore, one could make the case that the data used in Google Flu Trends is not from an empirical origin, like ASKAP. There are of course also similarities shared by these two systems. Leonelli argues that Mayer-Schönberger and Cukier consider three core innovations brought in by Big Data—present in the case of Google Flu Trends as well as ASKAP. The first one is *comprehensiveness*, and refers to the claim that accumulation of large data sets enables different scientists, at different times, to ground their analysis on different aspects the same phenomenon. Second comes *messiness*. Big Data pushes researchers to welcome the complex and multifaceted nature of the world, rather than pursuing exactitude and accuracy as standard scientific practice does. In fact, according to her, Mayer-Schönberger and Cukier's view takes that it is impossible to assemble Big Data in ways that are guaranteed to be accurate and homogeneous. Borrowing from the same idea, Mayer-Schönberger and Cukier say: "Big Data is messy, varies in quality, and is distributed across countless servers around the world" (Mayer-Schönberger and Cukier 2013, 13). Third and finally comes 'the triumph of *correlations*,' that is, a new kind of science (i.e., the science brought up by Big Data) guided by statistical correlations between data values rather than causality (Leonelli 2014).

This is a good place to stop and to ask the following question: what has Big Data to do with computer simulations? On the one hand, and as a general principle, they both share two common pillars. These are the *computational* pillar, which includes accounting for the speed of calculus, capacity to store data, and handling large volumes of data; and the *cognitive* pillar, which includes elucidating methods for representing, validating, aggregating, and cross-referencing data sets. On the other hand, Big Data and computer simulations have little in common. This is especially true when it comes to epistemic and methodological claims. We have already said something about the computational and cognitive pillars, for both Big Data and computer simulations. The question of interest now is, what are the differences that set Big Data and computer simulations apart. In other words, what make them two distinct technological paradigms?

6.3 The Fight for Causality: Big Data and Computer Simulations

A desirable aim for any scientific method is to be able to provide knowledge and understanding of the world that surrounds us. If this knowledge and understanding is delivered on a fast track, like in the case of Google Flue Trends, then it is certainly the most preferable. In any case, any such method needs to provide means for insight into the way the world is, otherwise it will be of little interest for scientific and engineering purposes. Take as an example our previous discussions about comparing computer simulations with laboratory experimentation in Chap. 3. Why do philosophers go into the trouble of discussing these issues? One chief reason is because we need to provide the means to warrant, within a reasonable scope, that the results of computer simulations are at least as reliable as our current best methods for inquiring into the world, that is, *via* experimentation. A similar reasoning applies for theory and models, as discussed in Chap. 2.

A core feature of experimentation, and one that much of the modeling practice also reflects, is that it uncovers causal relations. For instance, the antenna designed by Arno Penzias and Robert Wilson originally devised to detect radio waves ended up detecting cosmic microwave background radiation. How could this happen? The story is quite fascinating, but the important point for us is that such cosmic microwaves were causally interacting with the antenna. This, of course, occurred not without the scientists trying to eliminate other potential sources of causal interaction, such as terrestrial interference—noise emanating from New York City—as well as bat and dove droppings.

The reason why claims about causal relations are at the basis of the confidence of scientific and engineering methods is because they work as some sort of warrant for our knowledge and understanding of the world. Many philosophers have portrayed causal relations—or causality—as the cobble stones for any scientific enterprise. Philosopher John L. Mackie has even referred to causality as the 'cement of the universe' (Mackie 1980).

As the example of Penzias and Wilson shows, experimentation is crucial for science and engineering precisely because it helps to uncover the causal relations existing in the world. Let us recall from Chap. 3 the centrality of causality in discussions about the experimental membership of computer simulations. We saw that if we can show the existence of causal relations in computer simulations, then they can be treated on an equal footing as experiments. In case this is not possible, then their epistemological evaluation is still open. In Chap. 3 we had a relatively long discussion on the reasons that move some philosophers to believe that computer simulations lack causal relations, and what that means for their epistemological assessment. Whether causality comes from directly interacting with the world—such as in standard experimentation—or representing causes—such as in theories, models, and computer simulations—being able to identify causal relations is key for scientific and engineering progress. This is true, in any case, if we exclude Big Data. In Big Data, it has been argued, *correlation* supersedes *causality* in the race for knowing the world.

Mayer-Schönberger and Cukier are two influential authors that strongly preach for causation to be put into a drawer, and for researchers to embrace correlation as the new goal of scientific research instead. To them, it is utterly futile to direct efforts in discovering a universal mechanism for inferring causal relations from Big Data. At the end of the day, all we need for being epistemically satisfied in Big Data are high correlations (Mayer-Schönberger and Cukier 2013).

One question that we want to answer here is whether Big Data, as Mayer-Schönberger and Cukier depict it, could do well with just correlation, or if causality needs to be put back in the general picture of scientific and engineering research. My belief is that we shouldn't give up on causality so easily, mainly because doing so comes with a (high) price. Let us begin by providing some clarifications.

The basic difference between causality and correlation is that the former is taken as a regularity in a natural and social world that is stable and permanent throughout time and space. Correlation, on the other hand, is a statistical concept that measures the relationship between two given data values. For instance, there are no existing causal relations between playing football and breaking windows, despite the high correlation that we had in my childhood neighborhood. There is, however, a causal link between a football hitting my neighbor's window and the latter breaking into pieces. Causality, as we understand the term here, tells us the true reasons why and how something happens. Correlation, instead, only tells us something about a statistical relationships between two or more data points.

From the previous example we should not infer that correlation is not a secure source of knowledge. Studies on car insurance, for instance, have shown that male drivers have a stronger correlation with car accidents than female drivers, and therefore insurance companies tend to charge men more than women for their insurance. There are, of course, no causal relations between being a male driver and provoking an accident, nor being a female driver and being a safe driver. High correlation is, for cases like this, good enough.

Now, is this what Mayer-Schönberger and Cukier have in mind when they plead for correlation over causality in Big Data? Not quite. According to these authors, seeking causal relationships—a very hard and demanding task by itself—loses all its appeal with the exceptional increase of the available data about a given phenomenon. Recall from the previous section when we discussed 'comprehensiveness' and 'messiness' of data. While the former refers to the sheer amount of data, the latter emphasizes the role of inexact data. Together, Mayer-Schönberger and Cukier claim, they give the necessary epistemic backdrop to prefer high correlation over causation. How is this possible? The reasoning goes as follows: correlation consists in measuring the statistical relation between two data values, and data is precisely what researchers have in excess (i.e., the comprehensiveness thesis.[11]) This means that researchers are able to find very high and stable correlations in a dataset in a relatively easy and fast way. Moreover, researchers know that correlation could still remain high even if there is some inexact data, since the sheer quantity is sufficient for compensating any inexactitudes found in the data (i.e., the messiness thesis). Causality, then, becomes an antiquity under the new sun of Big Data, for researchers no longer need to find the acting causal relations in order to have claims about a given phenomenon. The example that illuminates this reasoning is Google Flu Trends again. As presented, Google engineers were able to find high correlations among the data despite the fact that some queries came from people experiencing regular flu, and some not being sick at all. The large amount of data gathered compensates for any inexactitudes hiding in these few queries. Naturally, the messiness thesis that allows us to disregard the few misleading queries presupposes that the percentage of such queries is negligible. Otherwise, correlation can still be high but about the wrong results.

As I mentioned earlier, Mayer-Schönberger and Cukier are right about the advantages of supporting, under the right circumstances, correlation over causation. And they profusely provide examples of successful cases. But I also warned that their ideas need to be in the proper context in order to be applicable. Most of the authors' examples come from the use of Big Data in large corporations and big government. Although the advantage that Big Data represents for these spheres of contemporary society is undeniable, they do not represent the uses and needs of Big Data in scientific and engineering contexts. Indeed, correlation seems to be the right conceptual tool for finding the relevant links among our buying preferences with future choices—as it is depicted in the case of Amazon.com (101)—and our financial credit scores with our personal behavior—as The Fair Isaac Corporation shows with the use of Big Data (56). But when the question is about the uses of Big Data for scientific and engineering purposes, causation seems to regain its original value. Let me explain why.

One of the major losses in trading correlation over causation is that the former cannot tell researchers precisely *why* something happens, but rather *what*—or *that*—it happens. Mayer-Schönberger and Cukier admit this from the beginning, putting these ideas in the following form: "if we can save money by knowing the best time to

---

[11]There is even the claim that the amount of data corresponds to the phenomenon itself. That is, the data *is* the phenomenon (Mayer-Schönberger and Cukier 2013).

buy a plane ticket without understanding the method behind airfare madness, that's good enough. Big data is about *what* not *why*" (Mayer-Schönberger and Cukier 2013, 14—Emphasis in original).[12]

According to Mayer-Schönberger and Cukier, then, Big Data is not in the business of providing explanation of why something is the case, but rather offering predictive and descriptive knowledge. This could be seen as a serious limitation for a method promoted as the fourth paradigm of scientific research. Physics without the ability to explain why the planets revolve around the sun would not advance our understanding of the solar system. Likewise, what separates evolutionists from creationists—and their cousins, the advocates of the intelligent design—is precisely their ability to explain much of our surrounding biological and geological world. A good example of evolution is the color variation of the peppered moth population as a consequence of the Industrial Revolution. In the pre-industrial era, the vast majority of these moths had a light, mottled coloring which worked as their camouflage against predators. It is estimated that before the industrial revolution, a uniformly dark variant of the peppered moth made up 2% of the species. After the industrial revolution, the color of the population experienced a deep change: up to 95% showed a dark coloration. The best explanation as to why this change took place stems from accounting for the adaptation of the moths to the new environment darked by pollution. The example stands as a major shift caused by mutations in a species, grounding the ideas of variation and natural selection. The epistemological power of evolutionary theory resides in the fact that can explain *why* this happened, and not just *that* it happened.

Losing explanatory force seems to be a too-high price to pay for any method that self-proclaims to be scientific. We have discussed in Sect. 5.1.1 how explanation could count as a genuine epistemological function of computer simulations, to the extent of grounding them as reliable methods for knowing and understanding the world. This is, of course, not to say that Big Data should be abandoned as a genuine method for seeking knowledge of the world. It is only meant to call attention to the limitations that Big Data has as a method for scientific and engineering research.[13]

Consider yet another case, this time coming from engineering. On the 28th of January, 1986, the Space Shuttle Challenger broke apart 73 s into its flight to finally disintegrate over the Atlantic Ocean. As a result, the disaster took the life of seven crew members. Richard Feynman, during the Challenger's hearing, showed that the rubbers used in the shuttle's O-rings became less resilient in situations where the ring was exposed to low temperatures. Feynman demonstrated that by compressing a sample of the O-rings in a clamp and immersing it in ice-cold water, the ring would never recover its original shape. With this evidence in place, the commission could

---

[12]The authors are of course aware of how important causality is in the sciences and engineering. In this respect, they say: "We will still need causal studies and controlled experiments with carefully curated data in certain cases, such as designing a critical airplane part. But for many everyday needs, knowing *what* not *why* is enough" (Mayer-Schönberger and Cukier 2013, 191—Emphasis in original).

[13]One could make the case that causality is not necessary for explanation, and that given the right explanatory framework, Big Data could be able to provide genuine explanations. To the best of my knowledge, such explanatory framework is still missing.

determine that the disaster was caused by the primary O-ring not properly sealing in the unusually cold weather of January 1986.

This second example shows that finding the right causal links, and thus being able to provide a genuine explanation, is not a matter that only concerns science, but technology as well. We keep in high esteem science and technology precisely because of their persuasive force for understanding and changing the world. Finding the right causal relations is one main contributor to this end. So, when Mayer-Schönberger and Cukier object that pinning down causal relations is a difficult task, this should not be taken as a reason for not going into their hunt. And when Jim Gray orders Big Data as the fourth paradigm of research, he encourage us to embrace its novelty as well as comprehend its limitations. No computational method in scientific and engineering research is all powerful that could dispense researchers from relying on traditional theorizing and experimenting.

On the other side, the notion of causality—or causal relations—is indeed hard to pin down. One might be tempted to suppose that all true regularities are causal relations. Unfortunately this is not the case. Many regularities that are familiar to us bare no causal relations whatsoever. The night follows the day as the day follows the night, but neither is causing the other. Moreover, one might assume that all laws of nature are somehow causal. But here also lies another disappointment. Kepler's law of planetary motion describes the orbit of the planets, but they offer no causal account of these motions. Similarly, the ideal gas law relates pressure with volume and temperature. It even tells us about how these quantities vary as functions of one another for a given sample of gas, but it tells us nothing about the causal relations existing among them.

Philosophers of science have been intrigued with causality for centuries. Aristotle is perhaps the most ancient and famous philosopher to discuss the notion and nature of causality (Falcon 2015). David Hume, on the other hand, was famous for having a skeptical view of causality, one that takes it to be a 'custom' or 'habit' that produces an idea of necessary connection (De Pierris and Friedman 2013). There are of course, a host of different philosophical interpretations of causality, reaching from ancient history to today.

Perhaps the most intuitive way to interpret causality is to take it as having some kind of *physical* feature. Take the example of the football hitting the window again. In this example, the ball physically interacts with the window, breaking it. To the proponents of physical causality, two objects are causally related when there is an exchange of a *physical quantity*, such as a *mark* or *momentum* from one onto the other.[14] The ball had an exchange of a physical quantity with the window, thus modifying it.

Unfortunately, not all causal relations in science and engineering depend in some way on an exchange of a physical quantity. Consider, for instance, the following sentence: "absenteeism in children at school age is *caused* by parents that have lost their jobs." In such a case, it is hard—if not impossible—to establish any physical causal relation between absenteeism and losing your job. For such cases, philosophers

---

[14]Prominent examples are Salmon (1998), Dowe (2000), and Bunge (1979).

tend to abstract from the 'physicality' of the causal relations, and put the emphasis on a representational level (Woodward 2003). It is this latter interpretation what enables computer simulations to embrace the notion of causality and Big Data to fight it.

Is it possible, then, to rescue causality in Big Data and thus keep our general view of scientific practice intact? In order to answer this question, we must first make clear how causality is found in these two technological paradigms. Causal models do exist in scientific practice, and they are regularly implemented as computer systems. The general characteristics of these models is that they accurately represent, under specific conditions, the set of causal relations playing a role in the target system.[15]

However, this is not the significance which we are giving here. For questions about causality to make sense to both computer simulations and Big Data, we need somehow to put them on an equal footing. Since Big Data is not about implementing models, but rather doing something with large amounts of data, then the only way to speak of causal relations relevant for both paradigms is when we take the causal relations *inferred* from the data. For such a case, an algorithm is in charged of reconstructing such causal relations from large amounts of data. In the following, I focus only on discussions about algorithms for Big Data, but what is being said here could also apply to computer simulations, provided that the reduction in the amount of data is being taken into account. In this context, the only difference between computer simulations and Big Data is on the amount of data that each one handles.

In the early 1990s, there was a fierce interest in entrenching the discovery of causal relations in data. At its core, the problem consists in having an algorithm that correctly identifies causal relations existing in large amounts of data. To illustrate how researchers typically know about causal relations, consider the case of smoking as a detriment to human health. Medical studies show that of the more than 7,000 chemicals in tobacco smoke, at least 250 are known to be harmful. And of those 250, at least 69 can cause cancer. These cancer-causing chemicals include acetaldehyde, aromatic amines, arsenic, etc. (U.S. Department of Health and Human Services 2014). How is such information typically obtained? The National Cancer Institute reports that their results are based on the available evidence, obtained by running standard medical trials.[16] In such a case, researchers have at their disposal a well populated corpus of scientific knowledge along with well established experimental methods from which they infer reliable information about chemicals that cause harm to human health. Consider again Google Flu Trends. Could researchers, doctors, and public health authorities infer existing causal relations that link the strain of the H1N1/09 with a large set of data? More specifically, the question that is at stake is whether there is an algorithm that could recreate causal relations existing between influenza A and the queries from data? This question actually keeps scientists and

---

[15]For a complete account of this interpretation of causality, see (Pearl 2000).

[16]It is interesting to note that The National Cancer Institute features four categories of causal relations, dependent on the strength of available evidence. Those are 'level 1: evidence is sufficient to infer a causal relationship', 'level 2: evidence is suggestive but not sufficient to infer causal relationship', 'level 3: Evidence is inadequate to infer the presence or absence of a causal relationship (which encompasses evidence that is sparse, of poor quality, or conflicting)', and 'level 4: evidence is suggestive of no causal relationship' (U.S. Department of Health and Human Services 2014).

engineers on tenterhooks, for a major triumph of Big Data would be able to infer causal relations from data by non-experimental means. The answers to this question, however, has researchers and philosophers divided.

In Spirtes et al. (1993), (henceforth SGS) presented their view on how to infer causal relations from data in a book they called *Causation, Prediction, and Search*. There, they claimed to have found a method for discovering causal relations based on data and which would not require any subject-matter knowledge. This means that scientists and engineers would be able to infer causal structures from collected data by simply running a special algorithm, regardless of their previous knowledge on the data or its causal structure. As expected, such an algorithm turned out to be quite sophisticated, combining graph theory, statistics, philosophy, and computer science.

Of course, SGS's algorithm was not the first attempt to infer causality from data. Conventional methods include data mining, regression analysis, path models, and factor analysis, among other existing causal methods from contemporary econometrics and sociometry. However, SGS claimed that theirs was a superior method. SGS particularly criticize regression analysis on the basis that, in terms of its causal structure—its causal claims—is not testable because it entails no constraints in the data. More generally, SGS identify three issues pervading conventional methods. First, they incorrectly identify causal hypotheses, they exclude causal hypotheses, and finally they also include many hypotheses that hold no causal significance. Second, the specifications of the distribution typically force the use of numerical procedures that are limited by statistical or numerical reasons. And third, restrictions on the search in the data space typically output to a single hypothesis, failing in this way to output alternative hypotheses which might be valid given the same data space.

The fundamental difference between SGS's algorithm and conventional attempts is, then, that none of the latter actually allow the identification of the right causal structures. Rather, and at best, they are able to discover patterns of association, which do not necessarily hold a causal structure. For each one of these methods, the general strategy is unsatisfactory since the goal is not only to identify the estimated distribution, but also to identify the causal structure and to predict future causal results of the variables. The SGS algorithm, it is claimed, complies with all of this.

In this context, SGS developed an algorithm called *TETRAD*.[17] Now, for TETRAD to be successful, the authors needed to pay attention to three key elements. First, the idea of a causal system with sufficient precision for mathematical analysis needed to be made explicit. Second, it was important to generalize their view to capture a wide range of scientific practice in order to understand the possibilities and limitations for discovering causal structures. And third, they needed to characterize the probabilities predicted by a causal hypothesis, given an intervention to a value (Spirtes et al. 1993, 3).

As a result, SGS were able to offer an algorithm that relates causal structures with a directed graph along with the probabilities assigned to each vertex of the graph. Furthermore, the authors claimed that the algorithm concerns the discovery

---

[17]For an up-to-date version of the TETRAD algorithm, visit http://www.phil.cmu.edu/tetrad/current.html.

of causal structures in linear as well as non-linear systems, systems with and without feedback, cases in which membership in the observed samples is influenced by the variables under study, and a handful of other cases (Spirtes et al. 1993, 4). As for the details of the TETRAD algorithm, I suggest the interested reader to directly approach SGS's book and website.[18] Here is not the right place for the heavy probabilistic and mathematical machinery there included. A quick overview of the objections that SGS had to face, however, is in order.

A few years later after the publication of SGS's book, Paul Humphreys and David Freedman launched a series of objections aiming to show that SGS have not accomplished what they claimed (Humphreys 1997; Freedman and Humphreys 1999). The core of Humphreys and Freedman's objections have two sources. On the one hand, SGS's analysis is tailored to technicalities that do not reflect the meaning of causation. In fact, according to Humphreys and Freedman, causation is assumed in the algorithms but never provided: "direct causes can be represented by arrows when the data are faithful to the true causal graph that generates the data" (Freedman and Humphreys 1999, 116). In other words, causation is defined in terms of causation, with little value added. The second objection is grounded on the lack of a satisfactory treatment of real problems of statistical inference stemming from imperfect data.

The very brief discussion presented above over reconstructing causal relations from data makes evident one important point for us, namely, the existence of serious problems for identifying causal structures in computer-based science. As argued, neither data produced by a computer simulation and of which researchers hold no information about their causal structures, nor data from Big Data can easily be used for inferring causal relations.

Perhaps a more tailor-made solution is needed. In this respect, one of the few defenders of causality in Big Data is Wolfgang Pietsch, who in a recent article argued for the causal character of modeling in Big Data (Pietsch 2015). Pietsch introduces and analyses several algorithms, concluding that they are able to identify causal relevance on the basis of eliminative induction and a related difference-making account of causation. This means that a phenomenon is examined under the systematic variation of potentially relevant conditions in order to establish causal relevance—or causal irrelevance—of such conditions. According to Pietsch, these must be done relative to a certain context or background determined by further conditions (Pietsch 2015). The author, then, takes that causal relevance of a condition could be determined by the method of difference. At its basis, the method compares two instances of causal relations which differ only on condition α, and agree in all other circumstances. If, in one instance, condition α and phenomenon Θ are present, and in another instance both are absent, then α is causally relevant to Θ. One way to determine this is by holding the following counterfactual: If α had not occurred, Θ would not have occurred. Following our example in medicine, one could say that if smoking had not occurred, cancer would not have occurred.

---

[18]See http://www.phil.cmu.edu/projects/tetrad/.

I am of course simplifying the difference-making account of causation. As one gets into the details, it is possible to discover what this account has to offer as well as its limitations. The problems discussed above constitute just one example of the kind of difficulty shared by these two paradigms of scientific research. Interestingly, one could in fact find more differences that set computer simulations and Big Data apart. One of such differences largely agreed upon is that simulations examine the implications of a mathematical model, whereas Big Data seek to find structures in large data sets. This, in turn, help us to delineate as diametrically opposed the sources from which researchers gather data. That is, while computer simulations *produces* large amounts of data from a given model, Big Data *recreates* a structure about a give phenomenon from large data sets. A further difference is that the ability of each one to predict, confirm, and observe a given phenomenon is different. Whereas computer simulations typically ground these epistemic functions in the model assumptions, Big Data relies on the collection of methods and curation processes underlying the structure of data. Their methodological approach is, therefore, different.

The starting point of the two is, therefore, also different. While computer simulations implement and solve a simulation model, Big Data examines a collection of data. This leads to a final difference; namely, the nature of our inferences about the target system. In computer simulations one typically derives consequences of the simulation model, whereas Dig Data is mostly driven by inductive inferences.

6.4 Concluding Remarks

When Jim Gray referred to computer simulations and Big Data as the third and fourth paradigm respectively he was, I believe, foreseeing the future of scientific and engineering research. In his last talk before he went missing at sea in 2007, he said "[t]he world of science has changed, and there is no question about this. The new model is for the data to be captured by instruments or generated by simulations before being processed by software and for the resulting information or knowledge to be stored in computers. Scientists only get to look at their data fairly late in this pipeline" (Gray 2009, xix).

Computer simulations have proven beyond doubt to be a fundamental tool for the advancement and development of scientific and engineering research. In this chapter, I have attempted to critically approach computer simulations and Big Data by showing their research agenda and how they differ in specific issues. I am aware that I have only scratched the surface of two methods with deep roots in the current—and future—state of science and engineering research. Nevertheless, let this chapter be a small contribution to a much broader and needed discussion.

Chapter 7
Ethics and Computer Simulations

This chapter has the sole purpose of asking the following question: is there an ethics that emerges in the context of computer simulations? In order to properly answer this question, we need to investigate the specialized literature to see how issues have been approached. The first problem that we encounter is the question of whether such an ethics actually exists, or rather if moral concerns in computer simulations can be approached by a more familiar ethical framework. Authors concerned with the general use of computers have already answered this question negatively, urging for a proper appraisal of computer ethics. It is interesting to note here the symmetry with our previous studies on the epistemology and methodology of computer simulations, particularly with the discussion presented in the introduction regarding their philosophical novelty. One could conjecture that it is part of the process of introducing new technology into scientific and engineering research that raises the eyebrow of the philosophically skeptical mind.

Unfortunately, very little work has been put into fully fleshing out an ethics to accommodate computer simulations. Only a handful of authors have addressed the above question at face value. In this chapter, I present and discuss them and their standpoints. In this respect, I will discuss concerns regarding the reliability of computer simulations—a topic that at this point is familiar to us—representation, and professional practice. Of these three approaches, only the last one is not directly dependent on the epistemology and methodology of computer simulation, and because of this I will dedicate some more time to discuss professional practice and the code of ethics for computer simulations in more detail. I will also reproduce and discuss the official code of ethics for researchers in computer simulations.

© Springer Nature Switzerland AG 2018

J. M. Durán, *Computer Simulations in Science and Engineering*,
The Frontiers Collection, https://doi.org/10.1007/978-3-319-90882-3_7

7.1 Computer Ethics, Ethics in Engineering, and Ethics in Science

In the past thirty or so years there has been an increasing interest for understanding ethics in the context of computers. Such an interest stems from the pervasive place that computers have in our regular life, as well as their presence in scientific and engineering practice. "Computer ethics," as it is today known is the branch of applied ethics focused on issues raised by computers, had a difficult beginning and it struggled to be confirmed as a discipline on its own right.

One of the foundational article for computer ethics was written by James H. Moor and appeared in 1985 under the title "What is computer ethics?" (Moor 1985). At the time, researchers working on applied ethics had the firm belief that the introduction of new technologies, such as the computer, did not suggest *new* moral issues worth studying; rather older and more familiar ethical problems could be applied to these technologies equally well.[1] Some even mocked the idea of an ethics for computers by suggesting an ethics of washing machines, and an ethics of cars. It soon became clear that there are no real grounds for comparing appliances and means of transportation with computers.

But let us get back to computer ethics. Moor's challenge was twofold. On the one hand, he had to show in what specific respect the studies on ethics carried out at his time fell short of accounting for computers. On the other hand, he had to show how computers brought new and genuine ethical issues to the table. Moor's strategy was to analyze the nature and social impact of computer technology, and the corresponding formulation and justification of policies for the ethical use of such technology. His concern was mostly on the policy vacuum that computers seemed to imply, as it was often the case at the time that there were neither regulations nor policies for design, programing, and use of software and hardware.

To put Moor's ideas into perspective, there are indeed cases where the computer is simply an accessory to the moral problem, and therefore it could be addressed from the viewpoint of standard ethical theories on the individual and society. For instance, that it is wrong to steal a computer. In this respect, the analysis would not have been any different from stealing a washing machine or a car. But Moor had something else in mind. To him, problems in computer ethics arose because there were no ethical frameworks that would indicate how computer technology should be used, nor how users, manufacturers, programmers, and basically everyone involved in the development of software and hardware should behave. As he cogently puts it, "a central task of computer ethics is to determine what we should do in such cases, i.e., to formulate policies to guide our actions. Of course, some ethical situations confront us as individuals and some as a society. Computer ethics includes consideration of both personal and social policies for the ethical use of computer technology"

---

[1]The difference between ethics and morals is often formulated in terms that ethics are the science of moral, whereas morals are the practice of ethics. While morals are the collection of values and standards, ethics is the formal study and encoding of those standards. I shall not make that distinction here. Instead, I will refer to both concept indistinguishably.

(Moor 1985, 266). As it turns out, he foresaw many of the ethical questions that are at the center of an ethics for computer simulations (see Sect. 7.3.2).[2]

Thus were the humble beginnings of computer ethics. Moor mainly discussed the revolutionary presence of computers in our everyday lives, as individuals, institutions, governments and as society. But it was not until the publication of Deborah Johnson's "Computer Ethics" (Johnson 1985) that the topic attracted more visibility. Now considered a classic book, Johnson also contributed to the justification of the existence and uniqueness of computer ethics. Unlike Moor, to whom she acknowledges setting the basis on the issue, Johnson's main interests were focused on issues over the Internet, privacy, and proprietary rights. Take for instance understanding privacy in the era of the Internet. Johnson's example are companies such as Amazon, who claimed that they needed to store information about their customers in order to serve them better. The question here is whether sending information about new book releases is actually providing a service, as the company claimed back in the middle 1990s, or it was somehow an invasion of privacy, since, most likely, customers did not explicitly request such information. Thus understood, Johnson claims that "the burden of proof is on privacy advocates to show either that there is something harmful about information gathering and exchange or that there is some benefit to be gained from constraining information gathering" (Johnson 1985, 119). Problems regarding gathering large amounts of data for purposes other than 'providing a service' was obviously not on the table back in the 1990s. Today, the ethical questions that companies such as Amazon, Google and other raise are, in fact, whether they are shaping our interest in books, movies, music and, ultimately, everything that concerns our personality, including our political views.[3]

Studies on computer ethics, then, put their emphasis on questions about how an action using a computer may be morally good or bad respective of its motives, consequences, universality, and virtuous nature. Thus understood, these studies have a broad range of topics, from the use of computers in and for the government, to their use in our private lives. This means that they do not necessarily focus exclusively on scientific and engineering uses of computers, let alone specifically computer simulations. Let me explain this. Studies on privacy and proprietary rights, for instance, do have an impact on the use of computers in scientific and engineering contexts. Ownership is usually discussed as a mechanism to control data, which could easily include cases of data from medical and drug testing. An issue that emerges in this context is the distinction between academic and public data, on the one hand, and commercial data, on the other. This distinction allows researchers, individuals, or institutions to retain realistic expectations over the potential uses and implications of

---

[2] Another author that anticipated much of the contemporary discussion is McLeod (1986).

[3] Studies on Big-Data constitutes a new *locus classicus* on issues regarding the ethical implications of companies dealing with large amounts of data. For cases of successful uses of Big-Data, see the work of Mayer-Schönberger and Cukier (2013); Marr (2016); and for the philosophical treatment, see among others (Zwitter 2014; Béranger 2016; Mittelstadt and Floridi 2016a). We could also mention the many institutions—governmental, educational, and in the private sector—that are being created to understand computer ethics. An excellent example of this is the Oxford Internet Institute, part of the University of Oxford.

their data (Lupton 2014). Thus understood, ethical concerns about proprietary rights do have an impact on the use of computers in scientific and engineering practice. Our concerns here, however, are on the moral issues emerging exclusively from the use of computers in scientific and engineering contexts, rather than extrapolations from other fields or areas. An example of this is moral issues tailored to representation of computer simulations (Sect. 7.2.2). The concerns that emerge in this context are, arguably, exclusively on the uses of computer simulations.

At this point we must ask, is there a branch of computer ethics that reaches out to moral issues emerging in the context of computer simulations? Surprisingly, the number of articles published on ethics involving computer simulations is very small. As we shall see in the next section, I present what are arguably the three main arguments on ethics of computer simulations available in the specialized literature. The first one discuss the work of Terence J. Williamson, an engineer and architect that presents ethical issues of computer simulations by looking at their epistemological problems. In this respect, he nicely connects our past discussions on the epistemology and methodology of computer simulations with moral concerns. The second study comes from the philosopher and member of the International Society for Ethics and Information Technology Philip Brey. Brey makes the interesting case that a significant aspect of ethics of computer simulations comes from their capacity to (mis)represent a target system. Since such a (mis)representation could come in several forms, it follows that there are specific professional responsibilities attached to each one of them. Our third discussion comes from a larger body of literature written by Tuncer Ören, Andreas Tolk, and others. Ören is an engineer trained that worked extensively in establishing ethical issues in computer simulations. He was also chief consultant in the elaboration of the code of ethics for researchers and practitioners of computer simulations (see Sect. 7.3).

We know that computer simulations combine two main components, namely, computer science and general science—or engineering, depending on what is being simulated. This suggests that any study on the ethics of computer simulations must include these two—or three—elements. That is to say, issues stemming from ethics of computers as well as ethics in science and engineering fuel ethical studies on computer simulations. We have discussed computer ethics, presenting very briefly some of the chief concerns that emerge in that field. Before discussing in some detail current approaches to ethics of computer simulations, let us briefly say something about ethics in science and engineering.

Studies on ethics in science and engineering are concerned with different issues attached to the professional behavior of researchers, on how they carry out their work and experiments, and the consequences that such products and experiments have in society. Now, whereas much of the researcher's daily work can be carried out without any visible ethical implications, there are several provisions offered by ethical studies that reinforce good scientific and engineering practice, along with an ethical framework within which to place the implications of their practices. Key themes include the fabrication of data, usually happening when researchers want to propound more vigorously and with greater conviction a particular hypotheses. As examples we can mention faking skin transplants by Summerlin in 1974 and the origin of experiments on cancer by Spectorin 1980 and 1981. Similarly, the falsification of data occurs when results are altered to fit within the researcher's intended results. There have been reported cases of falsification of results by a junior researcher because she was under pressure to duplicate her adviser's results. An example of this is the falsification of results on insulin receptors by Soman from 1978 to 1980. Further issues stemming from ethics in science and engineering can be identified: plagiarism, misconduct/theft, clinical trials, etc. For a short but illuminative analysis of these issues, see (Spier 2012). The works of Adam Briggle and Cart Mitcham are also key for studies on the ethics of scientific research (Briggle and Mitcham 2012). In the following, I present and discuss some of these issues in the context of ethics of computer simulations.

7.2 An Overview of the Ethics in Computer Simulations

7.2.1 Williamson

Let me begin with Williamson (2010) whose work, although chronologically more recent than Ören's and Brey's, has the advantage of being conceptually closer to our most recent discussions on the reliability of computer simulations—see Chap. 4.

The chief motivation that guides Williamson's ethical concerns is that computer simulations could help to improve human life as well as to contribute to a sustainable environment, now and in the future. Despite such optimistic ideals, Williamson choses to discuss ethical problems from a negative perspective. More specifically, Williamson argues that epistemic claims about the reliability of computer simulations may lead to spurious impressions of accuracy—and thus legitimacy—leading ultimately to morally inappropriate uses of computer simulations, such as wrong decision making and erroneous allocations of resources. In other words, the epistemology of computer simulation allows questions about values and ethics. In this context, Williamson shows that a combination of epistemic concerns about accuracy/validity and a given criteria for trustworthiness are constitutive of moral problems of computer simulations. Let us first analyze them in turn and later see how they combine to give rise to Williamson's moral concerns.

According to Williamson, accuracy or validity of a simulation model is conceived as the degree to which the model corresponds to the real-world phenomena under scrutiny. In order to ground the accuracy of the simulation model, Williamson relies on the standard methodologies of *empirical validation*, "which compares simulated results with measured data in the real world", *analytical verification*, "which compares simulation output from a program, subroutine, algorithm or software object with results from a known analytical solution or a set of quasi-analytical solutions", and *intermodal comparison*, "which compares the output of one program with the results of another similar program(s)" (Williamson 2010, 404). Much of the discussion on this issue have been already addressed in Chap. 4.

Unlike accuracy or validation, which are essentially issues tailored to the simulation model, the criteria of trustworthiness is closely related to the purposes of a given problem. That is, it depends on asking the right questions for the correct use of the simulation. The example that illustrates the criteria of trustworthiness comes from computer simulations of environment performance in buildings. In such simulations, researchers always need to ask specific questions related to the optimal level of occupant thermal comfort and the minimal values for energy costs (Williamson 2010, 405). All these questions presuppose a more or less coherent structure of beliefs that are constitutive of the trustworthiness of the simulation. More specifically, Williamson suggests that the trustworthiness of computer simulations can be identified by asking four questions about relevance.

- Absence: should that kind of knowledge be absent (or present)?
- Confusion: is there a distortion in the definition of knowledge?
- Uncertainty: what degree of certainty is relevant?
- Inaccuracy: how accurate does knowledge need to be? Is it irrelevant because it is not accurate enough, or is it needlessly accurate? (Williamson 2010, 405).

With these ideas in hand, Williamson proposes the following framework for the appropriate understanding of the ethical concerns that emerge in the context of computer simulations.

7.2.1.1   Credibility (and Absence)

According to Williamson, the credibility of the use of a simulation stems from two sources. On the one hand, the credibility of the simulation is dependent on its accuracy or validity. Indeed, an inaccurate simulation cannot provide knowledge (see our discussion in Sect. 4.1), and therefore it becomes a source of all sorts of ethical concerns. In fact, according to Williamson there is a direct proportion between accuracy and credibility: the more accurate a computer simulation, the more credible it is.

On the other hand, credibility depends on an authority capable of sanctioning the correlation between the simulation model—and its results—and the real-world (Williamson 2010, 406). In this respect, Williamson says "credibility will be lacking if moral responsibility (if not legal responsibility) in applying the results of simulations is not accepted at all levels" (Williamson 2010, 406). By 'all levels' he refers to the role of the expert sanctioning the reliability of the simulation. It is interesting to note how Williamson makes a clear distinction between the accuracy provided by specific, non-expert dependent methods (such as verification and validation methods), from the expert opinion. While most authors in the literature do not account for experts as reliable sources for sanctioning computer simulations, Williamson is uncomfortable with restricting credibility only to mathematical methods. He insists that the expert plays a fundamental role in the simulation to the point that "reflective judges [...] agree [what] should be included" (Williamson 2010, 406).

Credibility is weakened, however, in two specific cases, namely, either when key elements of a problem are absent when the experts agree that they should be included in the simulation (Williamson 2010, 406), and when the simulation is used inappropriately (e.g., to represent systems that the simulation is not capable of representing).

Let us illustrate these points with the use of a simulation of thermal performance in buildings in Adelaide, Australia (Williamson 2010, 407). According to the example, simulations of greenhouse emissions produced by the heating and cooling system do not correlate to actual performances derived from several years of recorded end-use energy data. The reason for this is that the simulation does not account for the heating and/or cooling appliances found in the Nationwide House Energy Rating Scheme—NatHERS. If it were not for the expert that determines that the results are wrong, the simulation could have lead to a spurious impression of accuracy, and therefore of legitimacy. The consequences would have been severe harm to the environment and human health. Furthermore, the Australian government would have not been able to achieve a sustainable policy since a good estimate of the potential energy consumption and greenhouse gas emission is not part of the simulation.

7.2.1.2 Transferability (and Confusion)

Transferability is understood as the possibility of using the results of a computer simulation beyond the intended range of the simulation model. To Williamson, questions regarding transferability must include the extent to which authoritative knowledge should be merely scientific—tailored to the computer simulation—or other forms of knowledge must also be taken into account.

Moral issues in the context of transferability emerge when responsible builders and architects want to achieve standards imposed by the industry (e.g., Star Rating) with the belief that, by doing so, it will further reduce energy consumption and greenhouse gas emission. Because the Star Rating is calculated from a simulation of the aggregate heating and cooling loads taking into account certain assumptions about the building, the results cannot be easily transferable to any other construction, but rather only to those that comply with the assumptions initially built into the model. In this respect, misunderstanding the assumptions built into a given simulation, as well as expecting to transfer the results of one simulation into a different context might seriously distort the target system with the consequence of wrongly allocating resources without achieving the desired goal (Williamson 2010, 407).

7.2.1.3 Dependability (and Uncertainty)

To Williamson, several computer simulations become no longer dependable because their results, after a specific period of time, are uncertain with respect to a given target system. Such uncertainty, it must be said, is not the product of ignorance by the researcher, nor a lack of initial accuracy by the results of the simulation—or its results—but rather of the changing nature of the real-world phenomena being simulated. Consider the following example. It is a well-known fact that bulk thermal insulation deteriorates over periods of time. It is expected that Fiberglass insulation in an attic space will significantly compress and it will lose as much as 30% of its insulation efficiency within a decade or so (Williamson 2010, 407). The implications for energy consumption and environmental impact will of course be significant, leading to serious concerns about human health, the rational use of energetic resources, etc. (Jamieson 2008). To Williamson's mind, current studies on the ethics of computer simulations "do not take this or other known unknowns into account" (Williamson 2010, 407). I believe he is correct on this account.

7.2.1.4   Confirmability (and Inaccuracy)

Simulation models include a series of construction assumptions, conjectures, and even data that do not necessarily match the epistemological status of scientific and engineering formulations (e.g., surface resistance, sky emissivity and discharge coefficient are included in a simulation of thermal performance). Inevitably, these assumptions and conjectures lead to uncertainties or, as Williamson likes to call them "'inaccuracies' in the application of the simulation" (Williamson 2010, 407). A good example of these is climate models, where it is not unusual that different global circulation models report different results. This is mostly due to the assumptions built into each model as well as the data used to instantiate them. Williamson shows this with the example of 17 global circulation climate models simulating a time series of globally averaged surface warming as reported in the 2007 Fourth Assessment Report (AR4) of the International Panel on Climate Change (see Fig. 7.1) (Meehl et al. 2007, 763). According to him, not all the predictions made by these models can be right or likely equally wrong. To counteract this effect, the average of the results is used as the most probable value for the future state.

Naturally, the inaccuracies found in the results of computer simulations raise concerns, if not straightforwards objections, to the use of results for ethically sensitive matters. But researchers are well aware that complete certainty is virtually impossible. The challenge, then, is to find ways of balancing the lack of information—or

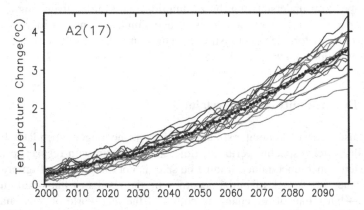

Fig. 7.1 Time series of globally averaged surface warming as reported in the 17 global circulation climate models, International Panel on Climate Change (IPCC) Scenario A2. Multi-model mean series are marked with black dots (Meehl et al. 2007, 763; Williamson 2010, 408)

unavoidable misinformation—that comes with certain kinds of computer simulations, and the ethical concerns that are pressing.

As we mentioned at the beginning, Williamson takes issue with the epistemic reliability of computer simulations, their failure as surrogates of reality, and the spurious impressions of legitimacy. One might not agree with him on the basis that all forms of knowledge (i.e., scientific, engineering, social, cultural, etc.) can be shown to be faulty, incomplete, or flawed. Does this mean that our use of it has negative moral consequences? The answer is a loud 'it depends'. For some cases, it is clearly true that incomplete knowledge leads to morally unacceptable consequences. The fabrication of data because of a lack of knowledge on how to correctly set up a scientific experiment has direct moral—and legal—consequences for researchers. And the use of fabricated data by politicians to inform the public is a serious matter.

The interesting work made by Williamson consists in showing how the epistemology of computer simulations extends its own domain and reaches out to value and ethical concerns, enable the legitimization of certain practices, and accounts for design decisions and regulations informed by simulations that are morally permissible. Because his preferred area of research is the architecture of buildings, his examples are related to environmental ethics. A good, final example is the development of building thermal performance simulations which enable, according to Williamson, an impressive growth in the selling of air-conditioning and the marketing of comfort conditions, both practices which are environmentally irresponsible (Shove 2003).

7.2.2 Brey

A related source of ethical concerns comes from the representational capacity of computer simulations as well as their professional use. Brey has argued that there are good reasons to believe that representations are not morally neutral. He uses as examples misrepresentations and biased representations in computer simulations. Similarly, Brey argues that the manufacture and use of computer simulations involve ethical choices, and therefore raises questions about professional practice (Brey 2008, 369).[4] Let us discuss these issues in order.

According to Brey, computer simulations may cause harm, as well as mislead researchers to the point of causing harm, if the simulation fails to uphold certain *standards of representational accuracy*. When does such a situation occur? In many cases, it is part of the aim and functionality of a computer simulation to realistically depict aspects of the real-world. For example, in a biomechanical simulation of bone-implant-systems, it is paramount to realistically and accurately represent the human bone, otherwise any information obtained from the simulation could be useless (Schneider and Resch 2014).

---

[4]Let it be said that Brey's work also relates ethical issues with Virtual Reality as used in video games or non-scientific visualizations. Here I focus solely on his view of computer simulations.

A core issue to discuss now is the question of what does it mean for a computer simulation to be realistic?—or, if you prefer, what does it mean to uphold acceptable standards of representational accuracy? We learned earlier from Williamson that the accuracy of results of a computer simulation are paramount for moral assessments. Now, the way Williamson understands the notion of *accuracy* is equivalent to *numerical accuracy*. That is, the accuracy of results of a computer simulation is understood as the degree to which they correspond to values measured and observed in a real-world target system. Brey, instead, has a different interpretation of accuracy in mind. In his own words, "[accuracy standards] are standards that define the degree of freedom that exist in the depiction of a phenomenon, and that specify what kinds of features must be included in a representation for it to be accurate, what level of detail is required, and what kinds of idealizations are permitted" (Brey 2008, 369). In other words, to Brey accuracy is synonymous with 'realism.' An accurate model is a model that realistically depicts the target system, that is, that reduces the number of idealizations, abstractions and fictionalization of the model, at the same time that it increases the level of detail.

Taking accuracy in this sense, we must now ask what sort of ethical issues emerge with *mis*representations? According to Brey, misrepresentations of reality are morally problematic to the extent that they can result in some form of harm. The greater these harms and the greater the chance of their occurrence, the greater the moral responsibility of the researchers. At this point it is interesting to note how Brey combines the misrepresentation of computer simulations—in principle morally neutral in and by itself—with the responsibility of the designers and manufacturers in ensuring the accuracy of representations (Brey 2008, 370). For instance, inaccuracies in reconstructing a simulated bone for a future bone implant can lead to grave consequences for the designers and manufacturers, but also for the patient. Another good example presented by Brey are the workings of an engine in educational software, where misrepresentations may lead students to hold false beliefs and, at a later point in time, cause some kind of harm (Brey 2008, 370).

Biased representations, on the other hand, constitute for Brey a second source of morally problematic representations in computer simulations (Brey 1999). A representation is biased when it selectively represents phenomena—or, we could add, selectively misrepresents phenomena. For example, it has been detected that in many global climate simulations the amount of carbon dioxide that plants absorb has been underestimated by about a sixth. This explains why CO_2 accumulation in the atmosphere is not as fast as climate models predicted it should be (Sun et al. 2014). Such bias representation unjustifiably ignores the contribution to global warming made by certain types of industries and countries. In general, the problems with biased representations is that they typically lead to unfair disadvantages for specific individuals or groups, or unjustifiably promote certain values or interests over others (Brey 2008, 370).

Furthermore, representations can be biased by containing implicit assumptions about the system in question. We already presented an example of implicit assumptions that lead to bias representations in our discussion on algorithms in Sect. 2.2.1.2—although in a different context. The example was to imagine the specification for a

simulation of a voting system. For this simulation to be successful, statistical modules are implemented in such a way that will give a reasonable distribution of the
voting population. During the specification stage, researchers decide to give more
statistical relevance to variables such as sex, gender, and health over other variables,
like education and income. If this design decision is not programed into the statistical module appropriately, then the simulation will never reflect the value of these
variables, thus resulting in a bias in the representation of the voters.

The example of the voting system brings forward some worries shared by many
authors, including Brey, Williamson and, as we shall see later, Ören. That is, the role
of the designers in the specification and programing of the computer simulations.
Brey argues that designers of computer simulations have the responsibility to reflect
on their values and ideals, potential bias and misrepresentations included in their
simulations, and to ensure that they do not violate important ethical principles (Brey
1999). To this end Brey, along with Williamson and Ören, resort to principles of professional practice and codes of ethics. As we shall see next, Ören not only presents
his concerns regarding the use of computer simulations, but he also proposes a solution through the implementation of a strict code of ethics exclusively for researchers
working with computer simulations.

7.2.3 Ören

Tuncer Ören has a more elaborate and thorough account of ethics in computer simulations. To his mind, the fundamental problem is whether the use of computer
simulations lead to serious implications for human beings. His chief concern stems
from the fact that computer simulations are used to support crucial policies and decisions that might alter our current life and constrain our future. In nuclear fuel waste
management systems, for instance, computer simulations are used to study the long
term behavior, environmental and social impact, as well as the means for containing
nuclear fuel waste. Similarly, computer simulations of leaking nuclear waste into
the Earth and underwater rivers are crucial for supporting decisions on whether such
nuclear waste should be buried instead of building a special facility for storage.

Ören of course understands that ethical questions regarding the use of results of
computer simulations have a direct relation to the reliability of computer simulations.
In this respect, he shares with Williamson and Brey the idea that the epistemology
of computer simulations is at the basis of moral concerns. About this, he says: "the
existence of several validation, verification, and accreditation techniques and tools
also attest to the importance of the implications of simulation" (Ören 2000). Similarly to Williamson and Brey, what it is important for Ören is to create the basis
for the credibility of computer simulations in such a way that decision and policy
makers can take simulations as a reliable tool. However, I believe that he goes further
than Williamson and Brey with his analysis when he claims that a proper study on
the ethics of computer simulations requires a well defined code of ethics that com-
plements verification, validation, and accreditation methods. To his mind, a code of
ethics would make it easier to establish the credibility of researchers designing and

programing the computer simulation—both as individuals and as groups—and thus not depend entirely on the computer simulation itself.[5]

An interesting issue emerging from Ören's position is that it depends heavily on the honesty, dedication, and proper conduct of the researchers.[6] This is of course a fair assumption. Scientific and engineering practice comes with the explicit adherence to the rules of good scientific practice, which includes maintaining professional standards, documenting results, rigorously questioning all findings, and honestly attributing any contributions by and to partners, competitors and predecessors (Deutsche Forschungsgemeinschaft 2013). National and international research foundations, universities and private funds, all understand that scientific misconduct is a serious offense to the scientific and engineering community, as well as poses a major threat to the prestige of any scientific institution. The German Research Foundation (DFG) takes scientific misconduct to be defined as "the intentional and grossly negligent statement of falsehoods in a scientific context, the violation of intellectual property rights or impeding on another person's research work" (Deutsche Forschungsgemeinschaft 2013). Any proved misconduct is seriously penalized. I shall discuss Ören's ideas on professional practice and code of ethics in the next sections. Before we get into that, let me introduce a small concern with his position.

The problem that I believe pervades Ören's position is that, if researchers adopt and live up to the rules of good scientific practice, then there is no special ethical problems for computer simulations. The reason for this is that, within Ören's framework, ethical issues are tailored to the researchers, and little accountability is put on the computer simulation itself. By this I of course do not mean to say that the computer simulation is morally responsible, but rather that ethical concerns stem from the simulation itself, as in the case made by Brey, where misrepresentation is a source of concern.

Ören is right, however, in pointing out how much researchers could benefit from having a code of ethics. Professional bodies have a long tradition in ascribing to such codes for the clear benefits they bring into their practice. Let me now discuss in more detail professional practice and the actual code of ethics for researchers working with computer simulations.

7.3 Professional Practice and a Code of Ethics

Codes of ethics originate in particular professional societies, such as the American Society of Civil Engineers, the American Society of Mechanical Engineers, the Institute of Electronic and Electrical Engineers, and others.[7] As suggested by their

---

[5]Let us recall that for Williamson, the 'expert' was not necessarily the researcher designing and programing a computer simulation, but rather the specialist on the topic that is being simulated.

[6]Andreas Tolk has a large body of work on the ethics of computer simulations that follows a similar line of research as Ören's. See Tolk (2017), and Armstrong and Taylor (2017).

[7]A collection of over 50 official codes of ethics issued by 45 associations in business, health, and law can be found in Gorlin (1994).

titles, these societies reflect the profession of their members—in these cases, all engineers—and their specialization.

Now, it is typical of such societies, groups, and institutions to subscribe to a *code of ethics*, that is, a collection of aspirations, regulations, and guidelines that represents the values of the profession to which it applies.[8] Such codes of ethics are customarily understood as normative standards about how engineers and scientists must conduct themselves in specific circumstances in order to remain part of the profession. In this sense, a code of ethics contains moral standards of what is understood, by a given community, as good and acceptable professional behavior.

If ethics in computer simulations depend on a code of ethics, as Ören suggests, then it will be paramount for us to be able to understand the functions of a code of ethics, its contents, and the applicability in studies on computer simulations.

Codes of ethics can be seen as a mark of a profession, and as such the provisions of the code of ethics are treated as guidelines that *ought* to be followed by the members of a society. One main function of the code of ethics is to contain a provision whereby a member should not do anything that could bring the profession to disrepute, or the professional to dishonor. This is a core element on which Ören's code of ethics depends, as it requires researchers' design and programing comply with the rules of good professional practice.

An important question here is whether a code of ethics presupposes a moral—and even legal—obligation for those that adhere to it. Some philosophers take it that a code of ethics is nothing more than an agreement among the members of a given profession to commit themselves to a common set of standards. Such standards serve to establish the basis and principles of a shared profession. Heinz Luegenbiehl, a philosopher interested in codes and education of engineers takes any code of ethics to be "a set of ethical rules that are to govern engineers in their professional lives" (Luegenbiehl 1991). Thus understood, a code of ethics only works to advise engineers on how they should act in given circumstances, but it enforces no moral or legal obligation, nor is it expected that engineers strictly comply with a code of ethics. In fact, under this interpretation, a code of ethics is nothing more than a commitment or duty to one's fellows professionals.

To other philosophers, a code of ethics is a set of statements embodying the collective wisdom of members of a given profession that tends to come to disuse. Practicing engineers seldom consult these codes, let alone follow them. There are several reasons for this. A prominent one is that researchers find that some of the code of ethics designed for their disciplines contain principles and ideals that are in conflict. Yet other philosophers find code of ethics coercive in intent, and thus limit the researcher's liberties as professionals. In fact, a basis for rejecting all codes of ethics stems from the claim that they presuppose a challenge of the autonomy of

---

[8]Codes come in many different forms. They can be formal (written) or informal (oral). They bear a variety of names, each with slightly different educational and regulatory aims. The most common forms are codes of ethics, codes of professional conduct, and codes of practice. Here, I will discuss a general form of codes. For more details, see Pritchard (1998).

moral agents. The question 'why do we need a code of ethics at all?' is answered by saying that it is perhaps because researchers are not moral agents.

Above and beyond the philosophical intricacies behind codes of ethics, and the reasons researchers have to adopt them or plainly reject them, there are several objectives that any code of ethics expects to fulfill. One important aim is to be inspirational, that is, it is supposed to inspire members to be more 'ethical' in their conduct. An obvious objection to this aim is that it presupposes that members of a professional community are unethical, amoral, or submoral, and thus it is necessary to exhort them—even by inspirational means—to be moral. Another important aim of codes of ethics is that they serve to sensitize members of a professional body to raise their concerns about moral issues related to their own discipline.

A code might also offer advice in cases of moral perplexity about what to do. For instance, when is it right for a member to report a colleague for wrongdoing? Naturally, such a case is different from a researcher violating a moral rule or even a law. Reporting a fellow colleague for selling medical drugs without the proper licenses and habitations should not constitute an issue for a code of ethics.

For the most part, a code of ethics works as a disciplinary code to enforce certain rules of a profession on its members in order to defend the integrity of the profession and protect its professional standards. It is doubtful, however, that any disciplinary action can be handled and enforced by a code of ethics, but it is, for the most part, a good source when used as criteria for determining malpractice.[9]

7.3.1   A Code of Ethics for Researchers in Computer Simulations

If by this point we are convinced of the value of ethics for designing, programing, and eventually using computer simulations, as Ören proposes, then the next question is how to develop a professional code of ethics for researchers working with computer simulations. Equally important will be to make explicit the responsibilities that lie above and beyond this code of ethics, which I also take from Ören.

The following is the *simulationist code of ethics* from the *The Society for Modeling and Simulation International (SCS)*. The SCS is dedicated to advancing the use and understanding of modeling and computer simulations with the purpose of solving real-world problems. It is interesting to note that, in their declaration, the SCS states their commitment not only to the advancement of computer simulations in areas of science and engineering, but also in the arts. Another important aim of the SCS is to promote communication and collaboration among professionals in the field of computer simulations. A chief reason for having a code of ethics is, precisely, to fulfill these constitutive objectives as a society.

---

[9]For more on the ethics of professional practice and codes of ethics, see Harris et al. (2009), and van der Poel and Royakkers (2011).

The code of ethics for researchers working in computer simulations, as published in the SCS's website (http://scs.org) is the following:

Simulationists are professionals involved in one or more of the following areas:

- Modeling and simulation activities.
- Providing modeling and simulation products.
- Providing modeling and simulation services.

1. Personal Development and Profession As a simulationist I will:

   1.1. Acquire and maintain professional competence and attitude.
   1.2. Treat fairly employees, clients, users, colleagues, and employers.
   1.3. Encourage and support new entrants to the profession.
   1.4. Support fellow practitioners and members of other professions who are engaged in modeling and simulation.
   1.5. Assist colleagues to achieve reliable results.
   1.6. Promote the reliable and credible use of modeling and simulation.
   1.7. Promote the modeling and simulation profession; e.g., advance public knowledge and appreciation of modeling and simulation and clarify and counter false or misleading statements.

2. Professional Competence As a simulationist I will:

   2.1. Assure product and/or service quality by the use of proper methodologies and technologies.
   2.2. Seek, utilize, and provide critical professional review.
   2.3. Recommend and stipulate proper and achievable goals for any project.
   2.4. Document simulation studies and/or systems comprehensibly and accurately to authorized parties.
   2.5. Provide full disclosure of system design assumptions and known limitations and problems to authorized parties.
   2.6. Be explicit and unequivocal about the conditions of applicability of specific models and associated simulation results.
   2.7. Caution against acceptance of modeling and simulation results when there is insufficient evidence of thorough validation and verification.
   2.8. Assure thorough and unbiased interpretations and evaluations of the results of modeling and simulation studies.

3. Trustworthiness As a simulationist I will:

   3.1. Be honest about any circumstances that might lead to conflict of interest.
   3.2. Honor contracts, agreements, and assigned responsibilities and accountabilities.
   3.3. Help develop an organizational environment that is supportive of ethical behavior.

3.4. Support studies which will not harm humans (current and future generations) as well as environment.
4. Property Rights and Due Credit As a simulationist I will:

   4.1. Give full acknowledgement to the contributions of others.
   4.2. Give proper credit for intellectual property.
   4.3. Honor property rights including copyrights and patents.
   4.4. Honor privacy rights of individuals and organizations as well as confidentiality of the relevant data and knowledge.
5. Compliance with the Code As a simulationist I will:

   5.1. Adhere to this code and encourage other simulationists to adhere to it.
   5.2. Treat violations of this code as inconsistent with being a simulationist.
   5.3. Seek advice from professional colleagues when faced with an ethical dilemma in modeling and simulation activities.
   5.4. Advise any professional society which supports this code of desirable updates.

The rationale for this code of ethics is provided by Ören in (2002). According to the author, there are at least two reasons for adopting a code of ethics for simulationists.[10] First, because there are a few emerging simulation societies that will require their members to adopt a code of ethics. In this way, and following the principles of a code of ethics, members can show the acceptance of their responsibility and accountability in developing, programing, and using computer simulations. Second, because computer simulations are a form of experimentation with dynamic models, and thus can affect humans and the environment in many diverse ways. In this context, ethicists need to provide an analysis of right and wrong actions, of good and bad consequences, and fair and unfair uses of computer simulations.

While the first reasons give by Ören aim at establishing criteria for professional responsibilities, the second furnishes a context for dealing with the consequences of the use of computer simulations.

7.3.2 Professional Responsibilities

The code of ethics presented above entrenches what constitutes good behavior for professional scientists and engineers working with computer simulations. In this respect, it focus attention on the design, programing, and use of computer simulations. In the following, I present Ören's discussion on the kind of responsibility ascribed to the professional practice of computer simulations. Following and extending (Ören 2000), scientists and engineers have a duty to the general public, their

---

[10]Ören actually speaks of three reasons, the third being the celebration of the 50th anniversary of the foundation of the Society for Modeling and Simulation International (SCS).

clients, the employers, colleagues, their profession, and themselves. The following list of responsibilities are taken from Ören (2000, 169).

- Responsibility with respect to the *public*:

  - A researcher shall act consistently with the public interest.
  - A researcher shall promote the use of simulation to better human existence.

- Responsibility with respect to *client* (a simulation client is a person, company, or agent which purchases, leases, or rents a simulation product, simulation service, or a simulation-based advice):

  - A researcher shall act in a manner that is in the best interest of the client. This responsibility shall be in accord with the public interest.
  - A researcher shall deliver/maintain a simulation product and/or services to solve the problem in the most trustworthy way. This includes usefulness, i.e., fitness for purpose(s) as well as meeting the highest professional standards.

- Responsibility with respect to the *employer*:

  - A researcher shall act in a manner that is in the best interest of the employer, provided that the activities are in alliance with the best interest of the public and the client.
    A researcher shall respect the intellectual property rights of her/his current and/or past employers.

- Responsibility with respect to the *colleagues*:

  - A researcher shall be fair and supportive of her/his colleagues.

- Responsibility with respect to the *profession*:

  - A researcher shall advance the integrity and reputation of the simulation profession consistent with the public interest.
  - A researcher shall apply simulation technology in the most appropriate way and shall not force simulation or any type of it as a Procrustean bed.
  - A researcher shall share her/his experience and knowledge to advance the simulation profession; and this will be in concert with her/his employers and clients interest.

- Responsibility with respect to *self*:

  - A researcher shall continue enhancing her/his abilities to have appropriate vision and knowledge to conceive problems from a broad perspective and to apply them for the solution of simulation problems.

Admittedly, there is little more to add to this code of ethics. On pain of repeating some of the issues here mentioned, this code of ethics could be extended on a few extra points. With respect to the responsibility towards a *client*, researchers must be very observant of all rules of confidentiality that relate their work with the interests of the client, as well as her ownership on designs, results, and the like. This could be the case of sensitive simulations such as in medicine. In such cases, the client trusts the researcher to keep the information provided in secrecy, a vote of trust that must not be broken.

In the case of the responsibility of a researcher with respect to the *public*, we could add that all public disclosure must be informed in the most objectively and unbiased way. As we discussed in Sect. 5.2.1, the visualization of a simulation is not straightforwardly clear for the non-expert (e.g., politician, public communicator), and therefore the communication of the visualization must not be mediated by subjective interpretations—as far as this is in fact possible. Although such responsibility is also extensible to clients and the employers, it is perhaps the general public the most sensitive agent to be protected from biased visualizations.

Finally, and as a general rule to observe, researchers must follow general codes of good professional conduct (e.g., the rules for the good scientific practice provided by the German Research Foundation (Deutsche Forschungsgemeinschaft 2013)). In this respect, scientists and engineers working with computer simulations do not have a special status as researchers, but rather they are subject to specific responsibilities tailored to their profession.

7.4   Concluding Remarks

Studies on the ethics of computer simulations are only in their infancy. In this chapter, I focused on presenting the main ideas regarding the ethics of computer simulations available in the current literature. In this respect, we saw three different standpoints. Williamson, who puts the emphasis on the epistemology and methodology of computer simulations, making clear that their reliability is paramount for the ethical assessment. Brey, who correctly calls attention to the dangers of misrepresentations and biased simulations. Finally, Ören who extensively discusses the form that professional responsibility should take in the context of computer simulations.

In a context where computer simulations are pervasive in scientific and engineering domains, where our knowledge and understanding of the inner workings of the world depends on them, it is striking that so little has been said about the moral consequences that follow from designing, programing, and using computer simulations. I hope that the discussion we have just had works as the starting point of many fruitful forthcoming discussions in ethics of computer simulations.

References

Ackermann, Robert. 1989. The New Experimentalism. *The British Journal for the Philosophy of Science* 40 (2): 185–190.

Ajelli, Marco, Hu Bruno Gonçalves, Duygu Balcan, Vittoria Colizza, José J. Hao, Stefano Merler Ramasco, and Alessandro Vespignani. 2010. Comparing Large-Scale Computational Approaches to Epidemic Modeling: Agent-Based versus Structured Metapopulation Models. *BMC Infectious Diseases* 10 (190): 1–13.

Alliance, International Virtual Observatory. 2018. http://www.ivoa.net. Accessed 26 Feb 2018.

Anthonie, Meijers. 2009. Philosophy of Technology and Engineering Sciences, Elsevier.

Appel, Kenneth, and Wolfgang Haken. 1976a. Every Planar Map is Four Colorable. *Bulletin of the American Mathematical Society* 82 (5): 711–712.

Appel, Kenneth, and Wolfgang Haken. 1976b. A Proof of the Four Color Theorem. *Discrete Mathematics* 16 (2): 179–180.

Ardourel, Vincent, and Julie Jebeile. 2017. On the Presumed Superiority of Analytical Solutions over Numerical Methods. *European Journal for Philosophy of Science* 7 (2): 201–220.

Aristotle. 1965. *Historia Animalium*. Harvard University Press.

Armstrong, Robert K., and Simon J. E. Taylor. 2017. Modeling and Simulation Societies Shaping the Profession. In *The Profession of Modeling and Simulation: Discipline, Ethics, Education, Vocation, Societies, and Economics*, ed. by Andreas Tolk and Tuncer Ören, 131–150. Wiley.

ASME. 2006. Guide for Verification and Validation in Computational Solid Mechanics. *The American Society of Mechanical Engineers, ASME Standard V&V* 10-2006.

Atkinson, Kendall E., Weimin Han, and David E. Stewart. 2009. *Numerical Solution of Ordinary Differential Equations*. Wiley

Bailer-Jones, Daniela. 2009. *Scientific Models in Philosophy of Science*. University of Pittsburgh Press.

Balcan, Duygu, Vittoria Colizza, Bruno Gonçalves, Hu Hao, José J. Ramasco, and Alessandro Vespignani. 2009. Multiscale Mobility Networks and the Spatial Spreading of Infectious Diseases. *Proceedings of the National Academy of Sciences of the United States of America* 106 (51): 21484–21489.

Banks, Jerry, John Carson, Barry L. Nelson, and David Nicol. 2010. *Discrete-Event System Simulation*. Prentice Hall.

Barberousse, Anouk, and Vorms Marion. 2013. Computer Simulations and Empirical Data. In *Computer Simulations and the Changing Face of Scientific Experimentation*, ed. by Juan M. Durán and Eckhart Arnold. 29–45 Cambridge Scholars Publishing.

© Springer Nature Switzerland AG 2018

J. M. Durán, *Computer Simulations in Science and Engineering*,

The Frontiers Collection, https://doi.org/10.1007/978-3-319-90882-3

Barberousse, Anouk, Sara Franceschelli, and Cyrille Imbert. 2009. Computer Simulations as Experiments. *Synthese* 169: 557–574.

Barker, Trish. 2004. Hunt for the supertwister. National Center for Supercomputing Applications—University of Illinois at Urbana-Campaign. http://www.ncsa.illinois.edu/news/stories/supertwister.

Barnes, Eric. 1994. Explaining Brute Facts. *Philosophy of Science* 1: 61–68.

Barnstorff, Kathy. 2010. X-51A Makes Longest Scramjet Flight. https://www.nasa.gov/topics/aeronautics/features/X-51A.html

Barrett, Jeffrey A., and P. Kyle Stanford. 2006. Prediction. In *The Philosophy, and of Science An Encyclopedia* ed. by S. Sarkar and J. Pfeifer, 585–599. Routledge.

Baumann, Robert. 2005. Soft Errors in Advanced Computer Systems. *IEEE Design Test of Computers* 22 (3): 258–266.

Beck, J.D., B.L. Canfield, S.M. Haddock, T.J.H. Chen, M. Kothari, and T.M. Keaveny. 1997. Three-Dimensional Imaging of Trabecular Bone Using the Computer Numerically Controlled Milling Technique. *Bone* 21 (3): 281–287.

Beisbart, Claus. 2012. How Can Computer Simulations Produce New Knowledge? *European Journal for Philosophy of Science* 2: 395–434.

Beisbart, Claus. 2017. Are computer simulations experiments? And if not, how are they related to each other? *European Journal for Philosophy of Science* 8(2): 171–204.

Bentsen, M., I. Bethke, J.B. Debernard, T. Iversen, A. Kirkevåg, Ø. Seland, H. Drange, et al. 2013. The Norwegian Earth System Model, NorESM1-M-Part 1: Description and Basic Evaluation of the Physical Climate. *Geoscientific Model Development* 6 (3): 687–720.

Béranger, Jérôme. 2016. *Big Data and Ethics: The Medical Datasphere*. Elsevier.

Berekovic, Mladen, Nikitas Simopoulos, and Stephan Wong (eds.). 2008. *Embedded Computer Systems: Architectures, Modeling, and Simulation*. Springer

Beyer, Mark, and Douglas Laney. 2012. The Importance of "Big Data": A Definition. Gartner, G00235055.

Bidoit, Michel, and Peter D. Mosses. 2004. *CASL User Manual: Introduction to Using the Common Algebraic Specification Language*. Springer.

Bird, Alexander. 2013. Thomas Kuhn. In *The Stanford Encyclopedia of Philosophy*, (Fall 2013 Edition), ed. by Edward N. Zalta. https://plato.stanford.edu/archives/fall2013/entries/thomas-kuhn/.

Birtwistle, Graham M. 1979. *DEMOS A System for Discrete Event Modelling on Simula*. The Macmillan Press. (Reprint 2003).

Bjorner, Dines, and Martin C. Henson. 2007. *Logics of Specification Languages*. Springer.

Blanco, Javier, and Pío García. 2011. A Categorial Mistake in the Formal Verification Debate. In *The Computational Turn: Past, Presents, Futures?*, ed. C. Ess, and R. Hagengruber. 30–33: Mv-Wissenschaft Århus University.

Blass, Andreas, and Yuri Gurevich. 2003. Algorithms: A Quest for Absolute Definitions. *In Bulletin of European Association for Theoretical Computer Science* 81: 195–225.

Blass, Andreas, Nachum Dershowitz, and Yuri Gurevich. 2009. When are Two Algorithms the Same? *The Bulletin of Symbolic Logic* 15 (2): 145–168.

Boehm, C., J.A. Schewtschenko, R.J. Wilkinson, C.M. Baugh, and S. Pascoli. 2014. Using the Milky Way Satellites to Study Interactions between Cold Dark Matter and Radiation. *Monthly Notices of the Royal Astronomical Society* 445: L31–L35.

Brewer, William F., and Bruce L. Lambert. 2001. The Theory-Ladenness of Observation and the Theory-Ladenness of the Rest of the Scientific Process. *Philosophy of Science* 68: S176–S186.

Brey, Philip. 1999. The Ethics of Representation and Action in Virtual Reality. *Ethics and Information Technology* 1 (1): 5–14.

Brey, Philip. 2008. Virtual Reality and Computer Simulation. *The Handbook of Information and Computer Ethics*, ed. by Kenneth Einar Himma and Herman T. Tavani, 361–384. Wiley.

Briggle, Adam. 2012. *Ethics and Science: An Introduction*. Cambridge: Cambridge University Press.

Buneman, Peter, James Cheney, Wang-Chiew Tan, and Stijn Vansummeren. 2008. Curated Databases. In *Proceedings of the Twenty-Seventh ACM SIGMODSIGACT- SIGART Symposium on Principles of Database Systems*, PODS '08, 1–12. ACM.

Bunge, Mario. 1979. *Causality and Modern Science*. Dover Publications.

Bunnik, Anno, Anthony Cawley, Michael Mulqueen, and Andrej Zwitter. 2016. *Big Data Challenges: Society, Security Innovation and Ethics*. Palgrave Macmillan.

Butcher, James. 2008. Modelling of Neuronal Circuits: The Future is Bright. *The Lancet Neurology* 7 (5): 382–383.

Cantwell-Smith, and Brian., 1985. The Limits of Correctness. *ACM SIGCAS Computers and Society* 14 (1): 18–26.

Callebaut, Werner. 2012. "Scientific perspectivism: A philosopher of science's response to the challenge of big data biology." *Studies in History and Philosophy of Science Part C: Studies in History and Philosophy of Biological and Biomedical Sciences* 43 (1): 69–80.

Ceruzzi, Paul E. 1998. *A History of Modern Computing*. MIT Press.

Chatrchyan, Serguei, V. Khachatryan, A.M. Sirunyan, A. Tumasyan, W. Adam, T. Bergauer, M. Dragicevic, J. Erö, C. Fabjan, M. Friedl, et al. 2014. Measurement of the Properties of a Higgs Boson in the Four-Lepton Final State. *Physical Review D* 89 (9): 1–75.

Chabert, Jean-Claude (ed.). 1994. *A History of Algorithms.*, From the Pebble to the Microchip. Springer.

Choudhury, Suparna, Jennifer R. Fishman, Michelle L. McGowan, and Eric T. Juengst. 2014. Big Data, Open Science and the Brain: Lessons Learned from Genomics. *Frontiers in Human Neuroscience* 8 (239). https://www.frontiersin.org/articles/10.3389/fnhum.2014.00239/full. Accessed 13 Dec 2017.

Ciletti, Michael D. 2010. *Advanced Digital Design with Verilog HDL*. Prentice Hall.

Cohn, Avra. 1989. The Notion of Proof in Hardware Verification. *Journal of Automated Reasoning* 5 (2): 127–139.

Colburn, Timothy, James H. Fetzer, and Terry L. Rankin (eds.). 1993. *Program Verification: Fundamental Issues in Computer Science*. Springer Science & Business Media.

Collins, Harry, and Robert Evans. 2007. *Rethinking Expertise*. University of Chicago Press.

Collmann, Jeff, and Sorin Adam Matei (eds.). 2016. *Ethical Reasoning in Big Data: An Exploratory Analysis*. Springer.

Copeland, B.Jack. 1996. What is Computation? *Synthese* 108 (3): 335–359.

Costa, Fabricio F. 2014. Big Data in Biomedicine. *Drug Discovery Today* 19 (4): 433–440.

Craver, Carl F. 2007. *Explaining the Brain: Mechanisms and the Mosaic Unity of Neuroscience*. Oxford University Press.

Craver, Carl F. 2001. Role Functions, Mechanisms, and Hierarchy. *Philosophy of Science* 68 (1): 53–74.

Critchlow, Terence, and Kerstin Kleese van Dam (eds.). 2013. *Data-Intensive Science*. Chapman & Hall/CRC.

Daston, Lorraine, and Peter Galison. 2007. *Objectivity*. Zone Books.

De Mol, and Liesbeth, and Giuseppe Primiero. 2014. Facing Computing as Technique: Towards a History and Philosophy of Computing. *Philosophy & Technology* 27 (3): 321–326.

De Mol, and Liesbeth, and Giuseppe Primiero. 2015. When Logic Meets Engineering: Introduction to Logical Issues in the History and Philosophy of Computer Science. *History and Philosophy of Logic* 36 (3): 195–204.

De Pierris, Graciela, and Michael Friedman. 2013. Kant and Hume on Causality. In *The Stanford Encyclopedia of Philosophy*, (Winter 2013 Edition), ed. by Edward N. Zalta. https://plato.stanford.edu/archives/win2013/entries/kant-hume-causality/.

Denzin, Norman K. (ed) 2006. *Sociological Methods: A Sourcebook*. Taylor and Francis.

Deutsche Forschungsgemeinschaft (ed.). 2013. *Sicherung Guter Wissenschaftlicher Praxis—Safeguarding Good Scientific Practice*. Wiley.

Dijkstra, Edsger W. 1974. Programming as a Discipline of Mathematical Nature. *American Mathematical Monthly* 81 (6): 608–612.

Douglas, Heather. 2009. *Science, Policy, and the Value-free Ideal*. University of Pittsburgh Press.

Douglas, Isbell, and Don Savage. 1999. *Mars Climate Orbiter Failure Board Releases Report, Numerous NASA Actions Underway in Response*. http://sunnyday.mit.edu/accidents/mco991110.html. Accessed 16 Aug 2018.

Dowe, Phil. 2000. *Physical Causation*. Cambridge University Press

Drogoul, Alexis, Jacques Ferber, and Christophe Cambier. 1992. Multi-agent Simulation as a Tool for Modeling Societies: Application to Social Differentiation in Ant Colonies. In *Simulating Societies*, ed. by G. Nigel Gilbert, 49–62. University of Surrey.

Durán, Juan M. 2013a. A Brief Overview of the Philosophical Study of Computer Simulations. *APA Newsletter on Philosophy and Computers* 13 (1): 38–46.

Durán, Juan M. 2013b. The Use of the 'Materiality Argument' in the Literature on Computer Simulations. In *Computer Simulations and the Changing Face of Scientific Experimentation*, ed. by Juan M. Durán and Eckhart Arnold, 76–98. Cambridge Scholars Publishing.

Durán, Juan M. 2014. Explaining Simulated Phenomena: A Defense of the Epistemic Power of Computer Simulations. *PhD dissertation, Universität Stuttgart*. https://elib.unistuttgart.de/handle/11682/5409.

Durán, Juan M. 2017a. Varieties of Simulations: From the Analogue to the Digital. In *Science and Art of Simulation 2015*, ed. by M. Resch, A. Kaminski, and P. Gehring. Springer.

Durán, Juan M. 2017b. Varying the Explanatory Span: Scientific Explanation for Computer Simulations. *International Studies in the Philosophy of Science* 31 (1): 27–45.

Durán, Juan M. and Nico Formanek 2018a. Essential Epistemic Opacity and Computational Reliabilism: Grounds for Trust. Unpublished.

Durán, Juan M. 2018b. Ciencia de la computación y filosofía: unidades de análisis del software. *Principia: An International Journal of Epistemology*, 22(2): 203–227.

Edwards, Kieran, and Mohamed Medhat Gaber (eds.). 2014. *Astronomy and Big Data. A Data Clustering Approach to Identifying Uncertain Galaxy Morphology* Cham: Springer.

El Skaf, and Rawad, and Cyrille Imbert. 2013. Unfolding in the Empirical Sciences: Experiments. *Thought Experiments and Computer Simulations. Synthese* 190 (16): 3451–3474.

Elgin, Catherine. 2007. Understanding and the Facts. *Philosophical Studies* 132 (1): 33–42.

Elgin, Catherine. 2009. Is Understanding Factive? In *Epistemic Value*, ed. by A. Haddock, A. Millar, and D.H. Pritchard, 322–330. Oxford University Press.

Fahrbach, Ludwig. 2005. Understanding Brute Facts. *Synthese* 145 (3): 449–466.

Falcon, Andrea. 2015. Aristotle on Causality. In *The Stanford Encyclopedia of Philosophy*, ed. by Edward N. Zalta. Spring 2015 Edition. https://plato.stanford.edu/archives/spr2015/entries/aristotlecausality/.

Fetzer, James H. 1988. Program Verification: The Very Idea. *Communications of the ACM* 37 (9): 1048–1063.

Feynman, Richard P. 2001. *What Do You Care What Other People Think?*. W.W: Norton & Company.

Floridi, Luciano. 2012. Big Data and Their Epistemological Challenge. *Philosophy and Technology* 25 (4): 435–437.

Floridi, Luciano, Nir Fresco, and Giuseppe Primiero. 2015. On Malfunctioning Software. *Synthese* 192 (4): 1199–1220.

Fox Keller, Evelyn. 2003. Models, Simulations, and "Computer Experiments". In *The Philosophy of Scientific Experimentation*, ed. by Hans Radder, 198–215. University of Pittsburgh Press.

Franklin, Allan. 1981. What Makes a 'Good' Experiment? *The British Journal for the Philosophy of Science* 32 (4): 367–374.

Franklin, Allan. 1986. *The Neglect of Experiment*. Cambridge University Press.

Freedman, David, and Paul W. Humphreys. 1999. Are There Algorithms That Discover Causal Structure? *Synthese* 121 (1): 29–54.

Fresco, Nir, and Giuseppe Primiero. 2013. Miscomputation. *Philosophy & Technology* 26 (3): 253–272.

Friedman, Michael. 1974. Explanation and Scientific Understanding. *The Journal of Philosophy* 71 (1): 5–19.

Frigg, Roman, and Stephan Hartmann. 2006. Scientific Models. In *The Philosophy of Science. An Encyclopedia*, ed. by S. Sarkar, and J. Pfeifer, 740–749. Routledge.

Frigg, Roman, and Julian Reiss. 2009. The Philosophy of Simulation: Hot New Issues or Same Old Stew? *Synthese* 169 (3): 593–613.

Fuchs, Norbert E. 1992. Specifications are (Preferably) Executable. *Software Engineering Journal* 7 (5): 323–334.

Galison, Peter. 1996. Computer Simulations and the Trading Zone. In *The Disunity of Science: Boundaries, Contexts, and Power*, ed. by Peter Galison and David J. Stump, 119–157. Stanford University Press.

García, Pío, and Marisa Velasco. 2013. Exploratory Strategies: Experiments and Simulations. In *Computer Simulations and the Changing Face of Scientific Experimentation*, ed. by Juan M. Durán and Eckhart Arnold. 99–117. Cambridge Scholars Publishing.

Gardner, Martin. 1970. The Fantastic Combinations of John Conway's New Solitaire Game "Life". *Scientific American* 223 (4): 120–123.

Gedenk, Eric. 2017. CAVE Visualization Room Immerses Researchers in Simulations. https:// www.hlrs.de/en/solutions-services/service-portfolio/visualization/virtual-reality/. Accessed 27 Jun 2016.

Gelfert, Axel. 2016. *How to Do Science with Models*. A Philosophical Primer. Springer Briefs in Philosophy: Springer.

Gelfert, Axel. 2018. Models in Search of Targets: Exploratory Modelling and the Case of Turing Patterns. In *Philosophy of Science*, ed. by Alexander Christian, David Hommen, Nina Retzlaff, and Gerhard Schurz. 9: 245–269. European Studies in Philosophy of Science, Springer, Cham.

Giere, Ronald N. 2009. Is Computer Simulation Changing the Face of Experimentation? *Philosophical Studies* 143 (1): 59–62.

Gilbert, Nigel, and Klaus G. Troitzsch. 2005. *Simulation for the Social Scientist*. 2nd ed. Open University Press.

Ginsberg, Jeremy, Matthew H. Mohebbi, Rajan S. Patel, Lynnette Brammer, Mark S. Smolinski, and Larry Brilliant. 2009. Detecting Influenza Epidemics Using Search Engine Query Data. *Nature* 457: 1012–1014.

Goldman, Alvin I. 1979. What is Justified Brief? In *Justification and Knowledge: New Studies in Epistemology*, ed. by George S. Pappas, 1–23. Springer.

Gorlin, Rena A. 1994. *Codes of Professional Responsibility*. BNA Books.

Gould, Harvey, Jan Tobochnik and Wolfgang Christian. 2007. *An Introduction to Computer Simulation Methods.*, Applications to Physical Systems. Pearson Addison Wesley.

Gray, Jim. 2009. Jim Gray on eScience: A Transformed Scientific Method. In *Fourth Paradigm Data-Intensive Scientific Discovery* ed by Tony Hey, Stewart Tansley, and Kristin Tolle, xvii–xxxi.

Grimm, Stephen R. 2010. The Goal of Explanation. *Studies in History and Philosophy of Science* 41: 337–344.

Guala, Francesco. 2002. Models, Simulations, and Experiments. In *Model-Based Reasoning: Science, Technology, Values*, ed. by Lorenzo Magnani and Nancy J. Nersessian, 59–74. Kluwer Academic.

Gupta, Anil K., Raj K. Singh, Sudheer Joseph, and Ellen Thomas. 2004. Indian Ocean High-Productivity Event (10–8 Ma): Linked to Global Cooling or to the Initiation of the Indian Monsoons? *Geology* 32 (9): 753–756.

Hacking, Ian. 1992. The Self-vindication of the Laboratory Sciences. In *Science as Practice and Culture*, ed. by Andrew Pickering, 29–64. University of Chicago Press.

Haddock, Adrian, Alan Millar, and Duncan Pritchard. 2009. *Epistemic Value*. Oxford University Press.

Halfhill, Tom R. 1995. The Truth Behind the Pentium Bug: An error in a lookup table created the infamous bug in Intel's latest processor. *BYTE* (March). https://web.archive.org/web/20060209005434/; http://www.byte.com/art/9503/sec13/art1.htm. Accessed 17 Oct 2017.

Halloran, M.Elizabeth, Neil M. Ferguson, Stephen Eubank, Ira M. Longini, Derek A.T. Cummings, Bryan Lewis, Xu Shufu, Christophe Fraser, Anil Vullikanti, Timothy C. Germann, et al. 2008. Modeling Targeted Layered Containment of an Influenza Pandemic in the United States. *Proceedings of the National Academy of Sciences*, ed. by Barry R. Bloom 105 (12): 4639–4644.

Hanson, Norwood Russell. 1958. *Patterns of Discovery: An Inquiry Into the Conceptual Foundations of Science*. Cambridge University Press.

Harris Jr, Charles E., Michael S. Pritchard, Michael J. Rabins. 2009. *Engineering Ethics Concepts and Cases*. Cengage Learning.

Hartmann, Stephan. 1996. The World as a Process: Simulations in the Natural and Social Sciences. In *Modelling and Simulation in the Social Sciences from the Philosophy of Science Point of View*, ed. by R. Hegselmann, Ulrich Mueller, and Klaus G. Troitzsch, 77–100. Kluwer.

Hartmann, Stephan. 1999. Models and Stories in Hadron Physics. In *Models as Mediators: Perspectives on Natural and Social Science*, ed. by Mary S. Morgan, and Margaret Morrison. 326–246 Cambridge University Press.

Hasse, Hans, and Johannes Lenhard. 2017. Boon and Bane: On the Role of Adjustable Parameters in Simulation Models. In *Mathematics as a Tool. Tracing New Roles of Mathematics in the Sciences*, ed. by Johannes Lenhard and M. Carrier. 93–116. Boston Studies in the Philosophy and History of Science. Springer.

Hempel, Carl G., and P. Oppenheim. 1948. Studies in the Logic of Explanation. *Philosophy of Science* 15 (2): 135–175.

Hempel, Carl G. 1965. *Aspects of Scientific Explanation and Other Essays in the Philosophy of Science*. The Free Press.

Hey, Tony, Stewart Tansley, and Kristin Tolle (eds.). 2009. *The Fourth Paradigm: Data-Intensive Scientific Discovery*. Microsoft Corporation.

Hick, Jason, Dick Watson, and Danny Cook. 2010. *HPSS in the Extreme Scale Era: Report to DOE Office of Science on HPSS in 2018–2022*. Report Number: LBNL–3877E.

Hoare, C.A.R. 1999. *A Theory Of Programming: Denotational, Algebraic and Operational Semantics*. Technical Report, Microsoft Research. https://www.microsoft.com/en-us/research/wpcontent/uploads/2016/02/a_theory_of_programming.pdf. Accessed 16 Aug 2018.

Hill, Robin K. 2013. What an Algorithm Is, and Is Not. *Communications of the ACM* 56 (6): 8–9.

Hill, Robin K. 2016. What an Algorithm Is. *Philosophy & Technology* 29 (1): 35–59.

Himeno, Ryutaro. 2013. Largest Neuronal Network Simulation Achieved Using K Computer. http://www.riken.jp/en/pr/press/2013/20130802_1/. Accessed 16 Aug 2018.

Hoare, C.A.R. 1971. Computer Science. In *Essays in Computing Science* (1989) ed. by C. B. Jones, 89–102. Prentice Hall International.

Hoose, Corinna, Jón Egill Kristjánsson, Jen-Ping Chen, and Anupam Hazra. 2010. A Classical-Theory-Based Parameterization of Heterogeneous Ice Nucleation by Mineral Dust, Soot, and Biological Particles in a Global Climate Model. *Journal of the Atmospheric Sciences* 67 (8): 2483–2503.

Humphreys, Paul, and Cyrille Imbert (eds.). 2012. *Models, Simulations, and Representations*. Routledge Studies in the Philosophy of Science: Routledge.

Humphreys, Paul W. 1990. PSA: Proceedings of the Biennial Meeting of the Philosophy of Science Association. *Computer Simulations* 2: 497–506.

Humphreys, Paul W. 1997. A Critical Appraisal of Causal Discovery Algorithms. In *Causality in Crisis?: Statistical Methods and the Search for Causal Knowledge in the Social Sciences*, ed. by Vaughn R. McKim and Stephen P. Turner, 249–263. University of Notre Dame Press.

Humphreys, Paul W. 2004. *Extending Ourselves: Computational Science, Empiricism, and Scientific Method*. Oxford University Press.

Humphreys, Paul W. 2009. The Philosophical Novelty of Computer Simulation Methods. *Synthese* 169 (3): 615–626.

Humphreys, Paul W. 2013a. Data Analysis: Models or Techniques? *Foundations of Science* 18 (3): 579–581.

Humphreys, Paul W. 2013b. What are Data About? In *Computer Simulations, and the Changing Face of Scientific Experimentation*, ed. by M. Juan. Durán and Eckhart Arnold, 12–28. Cambridge Scholars Publishing.

Ichikawa, Jonathan Jenkins, and Matthias Steup. 2012. The Analysis of Knowledge. In *The Stanford Encyclopedia of Philosophy*, Winter 2012, ed. by Edward N. Zalta. https://plato.stanford.edu/archives/win2012/entries/knowledge-analysis/. Accessed 16 Aug 2018.

IDC. 2014. The Digital Universe of Opportunities: Rich Data and the Increasing Value of the Internet of Things. https://www.emc.com/leadership/digital-universe/2014iview/executive-summary.htm. Accessed 11 Dec 2017.

Interstellar Medium of Isolated GAlaxies, AMIGA. 2018. http://amiga.iaa.es/p/1-homepage.htm. Accessed 26 Feb 2018.

Jamieson, Dale. 2008. *Ethics and the Environment: An Introduction*. Cambridge University Press.

Jason, Gary. 1989. The Role of Error in Computer Science. *Philosophia* 19 (4): 403–416.

Johnson, Deborah G. 1985. *Computer Ethics*. Englewood Cliffs.

Kaminski, Andreas, Michael Resch, and Uwe Küster. 2018. Mathematische Opazität. Über Rechfertigung und Reproduzierbarkeit in der Computersimulation. In *Jahrbuch Technikphilosophie*, ed. by Alexander Friedrich, Petra Gehring, Christoph Hubig, Andreas Kaminski, and Alfred Nordmann, 253–278. Nomos Verlagsgesellschaft.

Karaca, Koray. 2013. The Strong and Weak Senses of Theory-Ladenness of Experimentation: Theory-Driven versus Exploratory Experiments in the History of High-Energy Particle Physics. *Science in Context* 26 (1): 93–136.

Keaveny, Tony M., Edward F. Wachtel, X. Edward Guo, and Wilson C. Hayes. 1994. Mechanical Behavior of Damaged Trabecular Bone. *Journal of Biomechanics* 27 (11): 1309–1318.

Kennedy, Ashley Graham. 2012. A Non Representationalist View of Model Explanation. *Studies in History and Philosophy of Science Part A* 43 (2): 326–332.

Kennedy, Marc C., and Anthony O'Hagan. 2001. Bayesian Calibration of Computer Models. *Journal of the Royal Statistical Society: Series B (Statistical Methodology)* 63 (3): 425–464.

Kim, Jaegwon. 1994. Explanatory Knowledge and Metaphysical Dependence. *Philosophical Issues* 5: 51–69.

Kitcher, Philip. 1981. Explanatory Unification. *Philosophy of Science* 48 (4): 507–531.

Kitcher, Philip. 1989. Explanatory Unification and the Causal Structure of the World. In *Scientific Explanation*, ed. by Philip Kitcher, and Wesley C. Salmon, 410–505. University of Minnesota Press.

Kitchin, Rob. 2014. Big Data, New Epistemologies and Paradigm Shifts. *Big Data & Society* 1–12.

Knupp, Patrick, and Kambiz Salari. 2003. *Verification of Computer Codes in Computational Science and Engineering*. Chapman & Hall/CRC.

Knuth, Donald E. 1973. *The Art of Computer Programming*. Addison-Wesley.

Knuth, Donald E. 1974. Computer Science and Its Relation to Mathematics. *The American Mathematical Monthly* 81 (4): 323–343.

Knuuttila, Tarja. 2005. *Models as Epistemic Artefacts: Toward a Non-Representationalist Account of Scientific Representation*. Philosophical Studies from the University of Helsinki 8.

Kroepelin, S. 2006. Revisiting the Age of the Sahara Desert. *Science* 312(5777): 1138–1139

Krohs, Ulrich. 2008. How Digital Computer Simulations Explain Real-World Processes. *International Studies in the Philosophy of Science* 22 (3): 277–292.

Kuhn, Thomas S. 1962. *The Structure of Scientific Revolutions*. University of Chicago Press.

Kuhn, Thomas S. 1970. Logic of Discovery or Psychology of Research? In *Criticism and the Growth of Knowledge: Proceedings of the International Colloquium in the Philosophy of Science, London, 1965*, vol. ed. by Imre Lakatos and Alan Musgrave, 1–24. Cambridge University Press.

Küppers, Günter, and Johannes Lenhard. 2005. Validation of Simulation: Patterns in the Social and Natural Sciences. *Journal of Artificial Societies and Social Simulation* 8 (4). http://jasss.soc.surrey.ac.uk/8/4/3.html. Accessed 16 Aug 2018.

Laboratory, Los Alamos National. 2015. *Largest Computational Biology Simulation Mimics Life's Most Essential Nanomachine*. https://www.sciencedaily.com/releases/2005/11/051101223046.htm. Accessed 15 July 2017.

Laney, Doug. 2001. 3D Data Management: Controlling Data Volume, Velocity, and Variety. *Application Delivery Strategies* File 949. https://blogs.gartner.com/douglaney/files/2012/01/ad949-3D-Data-Management-Controlling-Data-Volume-Velocity-and-Variety.pdf. Accessed Feb 2017.

Latour, Bruno and Steve Woolgar. 1987. *Laboratory Life: The Construction of Scientific Facts*. Princeton University Press.

Lenhard, Johannes, and Eric Winsberg. 2010. Holism, Entrenchment, and the Future of Climate Model Pluralism. *Studies in History and Philosophy of Modern Physics* 41 (3): 253–262.

Lenhard, Johannes. 2006. Surprised by a Nanowire: Simulation, Control, and Understanding. *Philosophy of Science* 73 (5): 605–616.

Lenhard, Johannes. 2007. Computer Simulation: The Cooperation Between Experimenting and Modeling. *Philosophy of Science* 74: 176–194.

Leonelli, Sabina. 2014. What Difference Does Quantity Make? On the Epistemology of Big Data in Biology. *Big Data & Society* 1: 1–11.

Lesne, Annick. 2007. The Discrete Versus Continuous Controversy in Physics. *Mathematical Structures in Computer Science* 17 (2): 185–223.

Liberman, Mark. 2003. Zettascale Linguistics. http://itre.cis.upenn.edu/myl/languagelog/archives/000087.html. Accessed 10 March 2016.

Linder, C. 2012. A Numerical Investigation of the Exponential Electric Displacement Saturation Model in Fracturing Piezoelectric Ceramics. *Technische Mechanik* 32: 53–69.

Lipton, Peter. 2001. What Good is an Explanation? In *Explanation Theoretical Approaches and Applications*, ed. by Giora Hon and Sam S. Rakover. 43–59. Springer.

Lloyd, Elisabeth A. 1995. Objectivity and the Double Standard for Feminist Epistemologies. *Synthese* 104 (3): 351–381.

Longino, Helen E. 1990. *Science as Social Knowledge: Values and Objectivity in Scientific Inquiry*. Princeton University Press.

Luegenbiehl, Heinz. 1991. Codes of Ethics and the Moral Education of Engineers. In *Ethical Issues in Engineering*, ed. by Deborah Johnson, 4th edn, 137–138. Prentice Hall.

Lupton, Deborah. 2014. The Commodification of Patient Opinion: The Digital Patient Experience Economy in the Age of Big Data. *Sociology of Health & Illness* 36 (6): 856–869.

Machamer, Peter, Lindley Darden, and Carl F. Craver. 2000. Thinking about Mechanisms. *Philosophy of Science* 67 (1): 1–25.

MacKenzie, Donald A. 2001. *Mechanizing Proof: Computing, Risk, and Trust Inside Technology*. MIT Press.

Mackie, J.L. 1980. *The Cement of the Universe*. Oxford University Press.

Mahajan, Roop L., Rolf Mueller, Christopher B. Williams, Jeff Reed, Thomas A. Campbell, and Naren Ramakrishnan. 2012. Cultivating Emerging and Black Swan Technologies. *ASME 2012 International Mechanical Engineering Congress and Exposition* 6: 549–557.

Marr, Bernard. 2016. *Big Data in Practice: How 45 Successful Companies Used Big Data Analytics to Deliver Extraordinary Results*. Wiley.

Massimi, Michela, and Wahid Bhimji. 2015. Computer Simulations and Experiments: The Case of the Higgs Boson. *Studies in History and Philosophy of Science Part B: Studies in History and Philosophy of Modern Physics* 51: 71–81.

Mayer-Schönberger, Viktor, and Kenneth Cukier. 2013. *Big Data: A Revolution That Will Transform How We Live, Work, and Think*. Houghton Mifflin Harcourt.

Mayo, Deborah G., and Aris Spanos, eds. 2010. *Error and Inference. Recent Exchanges on Experimental Reasoning, Reliability, and the Objectivity and Rationality of Science*. Cambridge University Press.

Mayo, Deborah G. 1994. The New Experimentalism, Topical Hypotheses, and Learning from Error. *In PSA: Proceedings of the Biennial Meeting of the Philosophy of Science Association* 1: 270–279.

Mayo, Deborah G. 1996. *Error and the Growth of Experimental Knowledge*. University of Chicago Press.

Mayo, Deborah G. 2010. Learning from Error, Severe Testing, and the Growth of Theoretical Knowledge. In *Error and Inference. Recent Exchanges on Experimental Reasoning, Reliability, and the Objectivity and Rationality of Science*, ed. by Deborah G. Mayo and Aris Spanos. 28–57. University of Chicago Press.

McFarland, John, and Sankaran Mahadevan. 2008. Multivariate Significance Testing and Model Calibration under Uncertainty. *Computer Methods in Applied Mechanics and Engineering* 197: 2467–2479.

McLeod, John. 1986. But, Mr. President—Is it Ethical? In *Proceedings of the 1986 Winter Simulation Conference* ed. by J. Wilson, J. Henriksen, and S. Roberts. 69–71.

McMillan, Claude, and Richard F. González. 1965. *Systems Analysis: A Computer Approach to Decision Models*. Homewood/Ill: Irwin.

Meehl, Gerard A., Thomas F. Stocker, William D. Collins, A. T. Friedlingstein, T. Gaye Amadou, M. Gregory Jonathan, Akio Kitoh et al. 2007. Climate Change 2007: The Physical Science Basis. Contribution of Working Group I to the Fourth Assessment Report of the Intergovernmental Panel on Climate Change. In Chap. *Global Climate Projections*, ed. by S. Solomon, D. Qin, M. Manning, Z. Chen, M. Marquis, K.B. Averyt, M. Tignor, and H.L. Miller, 747–845. Cambridge University Press.

Millo, De, A. Richard, Richard J. Lipton, and Alan J. Perlis. 1979. Social Processes and Proofs of Theorems and Programs. *Communications of the ACM* 22 (5): 271–280.

Mittelstadt, Brent Daniel, and Luciano Floridi. 2016a. The Ethics of Big Data: Current and Foreseeable Issues in Biomedical Contexts. *Science and Engineering Ethics* 22 (2): 303–341.

Mittelstadt, Brent Daniel, and Luciano Floridi. 2016b. *The Ethics of Biomedical Big Data*. Springer.

Moor, James H. 1985. What Is Computer Ethics? *Metaphilosophy* 16 (4): 266–275.

Moor, James H. 1988. The Pseudorealization Fallacy and the Chinese Room Argument. In *Aspects of Artificial Intelligence*, ed. by James Fetzer, 35–53. Springer.

Morgan, Mary S., and Margaret Morrison (eds.). 1999. *Models as Mediators: Perspectives on Natural and Social Sciences*. Cambridge University Press.

Morgan, Mary S. 2003. Experiments without Material Intervention. In *The Philosophy of Scientific Experimentation*, ed. by Hans Radder, 216–235. University of Pittsburgh Press.

Morgan, Mary S. 2005. Experiments Versus Models: New Phenomena, Inference and Surprise. *Journal of Economic Methodology* 12 (2): 317–329.

Morrison, Margaret. 2009. Models, Measurement and Computer Simulation: The Changing Face of Experimentation. *Philosophical Studies* 143 (1): 33–57.

Morrison, Margaret. 2015. *Reconstructing Reality. Models, Mathematics, and Simulations*. Oxford University Press.

Müller, Tibor, and Harmund Müller. 2003. *Modelling in Natural Sciences*. Springer.

Naylor, Thomas H., William H. Wallace, and W. Earl Sasser. 1967a. A Computer Simulation Model of the Textile Industry. *Journal of the American Statistical Association* 62 (320): 1338–1364.

Naylor, Thomas H., J.M. Finger, James L. McKenney, Williams E. Schrank, and Charles C. Holt. 1967b. Verification of Computer Simulation Models. *Management Science* 14 (2): 92–106.

Newman, Julian. 2016. Epistemic Opacity, Confirmation Holism and Technical Debt: Computer Simulation in the Light of Empirical Software Engineering. In *History and Philosophy of Computing—Third International Conference, HaPoC 2015*, ed. by Fabio Gadducci and Mirko Tavosanis, 256–272. Springer.

Niebur, Glen L., Michael J. Feldstein, Jonathan C. Yuen, Tony J. Chen, and Tony M. Keaveny. 2000. High-Resolution Finite Element Models with Tissue Strength Asymmetry Accurately Predict Failure of Trabecular Bone. *Journal of Biomechanics* 33 (12): 1575–1583.

NSF. 2012. Core Techniques and Technologies for Advancing Big Data Science & Engineering (BIGDATA). NSF 12-499. https://www.nsf.gov/pubs/2012/nsf12499/nsf12499.htm. Accessed 17 Oct 2017.

NSF. 2014. Critical Techniques and Technologies for Advancing Big Data Science & Engineering (BIGDATA). NSF 14-543. https://www.nsf.gov/pubs/2014/nsf14543/nsf14543.htm. Accessed 17 Oct 2017.

NSF. 2016. Critical Techniques, Technologies and Methodologies for Advancing Foundations and Applications of Big Data Sciences and Engineering (BIGDATA). NSF 16-512. https://www.nsf.gov/pubs/2016/nsf16512/nsf16512.htm. Accessed 17 Oct 2017.

Oberkampf, William L., and Christopher J. Roy. 2010. *Verification and Validation in Scientific Computing*. Cambridge University Press.

Oberkampf, William L., and Timothy G. Trucano. 2002. Verification and Validation in Computational Fluid Dynamics. *Progress in Aerospace Sciences* 38 (3): 209–272.

Oberkampf, William L., and Timothy G. Trucano. 2008. Verification and Validation Benchmarks. *Nuclear Engineering and Design* 238 (3): 716–743.

Oberkampf, William L., Timothy G. Trucano, and Charles Hirsch. 2003. *Verification, Validation, and Predictive Capability in Computational Engineering and Physics*. Department of Energy: Sandia National Laboratories. United States.

Ören, Tuncer, Bernard P. Zeigler, and Maurice S. Elzas (eds.). 1982. In *Proceedings of the NATO Advanced Study Institute on Simulation and Mode-Based Methodologies: An Integrative View*. NATO ASI Series. Springer.

Ören, Tuncer. 1984. Model-Based Activities: A Paradigm Shift. In *Simulation and Model-Based Methodologies: An Integrative View*, ed. by Tuncer Ören, B.P Zeigler and M.S. Elzas, 3–40. Springer.

Ören, Tuncer I. 2000. Responsibility, Ethics, and Simulation. *Transactions of The Society for Computer Simulation International* 17 (4): 165–170.

Ören, Tuncer I. 2002. Rationale for a Code of Professional Ethics for Simulationists. In *Summer Computer Simulation Conference*, 428–433. Society for Computer Simulation International, 1998.

Ören, Tuncer. 2011a. A Critical Review of Definitions and About 400 Types of Modeling and Simulation. *SCS M&S Magazine* 2 (3): 142–151.

Ören, Tuncer. 2011b. The Many Facets of Simulation through a Collection of about 100 Definitions. *SCS M&S Magazine* 2 (2): 82–92.

Oreskes, Naomi, and Kristin. Shrader-Frechette, and Kenneth Belitz. 1994. Verification, Validation, and Confirmation of Numerical Models in the Earth Sciences. *Science* 263 (5147): 641–646.

Orf, Leigh, Robert Wilhelmson, Lou Wicker, B.D. Lee, and C.A. Finley. 2014. Talk 3B.3 Genesis and maintenance of a long-track EF5 tornado embedded within a simulated supercell the *27th Conference on Severe Local Storms*. https://ams.confex.com/ams/27SLS/webprogram/Paper255451.html. Accessed 21 September 2016.

Parke, Emily C. 2014. Experiments, Simulations, and Epistemic Privilege. *Philoophy of Science* 81 (4): 516–536.

Parker, Wendy S. 2008. Computer Simulations Throught an Error-Statistical Lens. *Synthese* 163 (3): 371–384.

Parker, Wendy S. 2009. Does Matter Really Matters? Computer Simulations, Experiments, and Materiality. *Synthese* 169 (3): 483–496.

Parker, Wendy S. 2014. Simulation and Understanding in the Study of Weather and Climate. *Perspectives on Science* 22 (3): 336–356.

Pearl, Judea. 2000. *Causality. Models, Reasoning, and Inference*. Cambridge University Press.

Perini, Laura. 2004. Visual Representations and Confirmation. *Philosophy of Science* 72 (5): 913–916.

Perini, Laura. 2005. The Truth in Pictures. *Philosophy of Science* 72 (1): 262–285.

Perini, Laura. 2006. Visual Representation. In: *The Philosophy of Science. An Encyclopedia*, ed. by Sahotra Sarkar and Jessica Pfeifer, 863–870. Routledge.

Pfleeger, Shari Lawrence, and Joanne M. Atlee. 2010. *Software Engineering: Theory and Practice*. Prentice Hall.

Piccinini, Gualtiero. 2007. Computing Mechanisms. *Philosophy of Science* 74: 501–526.

Piccinini, Gualtiero. 2008. Computation without Representation. *Philosophical Studies* 137 (2): 205–241.

Pietsch, Wolfgang. 2015. The Causal Nature of Modeling with Big Data. *Philosophy & Technology* 29 (2): 137–171.

Press, William H., Saul A. Teukolsky, William T. Vetterling, and Brian P. Flannery. 2007. *Numerical Recipes. The Art of Scientific Computing*. Cambridge University Press.

Primiero, Giuseppe. 2014. On the Ontology of the Computing Process and the Epistemology of the Computed. *Philosophy & Technology* 27 (3): 485–489.

Primiero, Giuseppe. 2016. Information in the Philosophy of Computer Science. In *The Routledge Handbook of Philosophy of Information*, ed. by Luciano Floridi, 90–106. Rouledge.

Pritchard, Duncan. 2010. *What is this Thing Called Knowledge?* Routledge.

Pritchard, J. 1998. Codes of Ethics. In *Encyclopedia of Applied Ethics (Second Edition)*, ed. by Ruth Chadwick, 494–499. Academic Press.

Radford, Colin. 1966. Knowledge by Examples. *Analysis* 27 (1): 1–11.

Ramstein, Gilles, Frédéric Fluteau, Jean Besse, and Sylvie Joussaume. 1997. Effect of Orogeny, Plate Motion and Land-Sea Distribution on Eurasian Climate Change over the Past 30 Million Years. *Nature* 386: 788–795.

Rapaport, William J. 1999. Implementation is Semantic Interpretation. *The Monist* 82 (1): 109–130.

Rapaport, William J. 2005. Implementation is Semantic Interpretation: Further Thoughts. *Journal of Experimental & Theoretical Artificial Intelligence* 17 (4): 385–417.

Rimoldi, Adele. 2011. Simulation Strategies for the ATLA Experiment at LHC. In *Journal of Physics: Conference Series*, 331: 1–8. IOP Publishing. http://inspirehep.net/record/1117099/files/jpconf11_331_032026.pdf. Accessed 12 July 2018.

Rohrlich, Fritz. 1990. Computer Simulation in the Physical Sciences. *Philosophy of Science Association* 2: 507–518.

Rojas, Raúl, and Ulf Hashagen (eds.). 2000. *The First Computers: History and Architectures*. MIT Press.

Saam, Nicole J. 2016. What is a Computer Simulation? A Review of a Passionate Debate. *Journal for General Philosophy of Science* 48 (2): 293–309.

Safran, Charles, Meryl Bloomrosen, W. Edward Hammond, Steven Labkoff, Suzanne Markel-Fox, Paul C. Tang, and Don E. Detmer. 2006. Toward a National Framework for the Secondary Use of Health Data: An American Medical Informatics Association White Paper. *Journal of the American Medical Informatics Association* 14 (1): 1–9.

Salmon, Wesley C. 1984. *Scientific Explanation and the Causal Structure of the World*. Princeton University Press.

Salmon, Wesley C. 1989. *Four Decades of Scientific Explanation*. University of Pittsburgh Press.

Salmon, Wesley C. 1998. *Causality and Explanation*. Oxford University Press.

Sargent, Robert G. 2007. Verification and Validation of Simulation Models. In *Proceedings of the 2007 Winter Simulation Conference* ed. by S. G. Henderson, B. Biller, M.-H. Hsieh, J. Shortle, J. D. Tew, and R. R. Barton, 124–137. IEEE Computer Society Press.

Schelling, Thomas C. 1971. On the Ecology of Micromotives. *National Affairs* 25 (Fall). https://www.nationalaffairs.com/public_interest/detail/on-the-ecology-of-micromotives. Accessed 12 Aug 2017.

Schiaffonati, Viola. 2016. Stretching the Traditional Notion of Experiment in Computing: Explorative Experiments. *Science and Engineering Ethics* 22 (3): 647–665.

Schneider, Ralf, and Michael M Resch. 2014. Calculation of the Discrete Effective Stiffness of Cancellous Bone by Direct Mechanical Simulations. In *Computational Surgery and Dual Training*, ed. by Marc Garbey, Barbara L. Bass, Scott Berceli, Christophe Collet, and Pietro Cerveri, 351–361. Springer.

Schurz, Gerhard, and Karel Lambert. 1994. Outline of a Theory of Scientific Understanding. *Synthese* 101: 65–120.

Schuster, Mathieu, Philippe Duringer, and Jean-Françpos Gaine, Patrick Vignaud, Hassan T. Mackaye, Andossa Likius, and Michel Brunet. 2006. The Age of the Sahara Desert. *Science* 311 (5762): 821–821.

Seibel, Peter. 2009. *Coders at Work: Reflections on the Craft of Programming*. Apress.

Shackley, Simon, Peter Young, Stuart Parkinson, and Brian Wynne. 1998. Uncertainty, Complexity and Concepts of Good Science in Climate Change Modelling: Are GCMs the Best Tools? *Climatic Change* 38 (2): 159–205.

Shannon, Robert E. 1975. *Systems Simulation: The Art and Science*. Prentice Hall.

Shannon, Robert E. 1978. Design and Analysis of Simulation Experiments. In *Proceedings of the 10th Conference on Winter Simulation*, vol. I, 53–61. IEEE Press.

Shannon, Robert E. 1998. Introduction to the Art and Science of Simulation. In *Proceedings of the 1998 Winter Simulation Conference*, ed. by , D.J. Medeiros, E.F. Watson, J.S. Carson and M.S. Manivannan, 7–14. Los Alamitos: IEEE Computer Society Press.

Shove, Elizabeth. 2003. *Comfort. Cleanliness and Convenience: The Social Organization of Normality*. Berg Publishers.

Shubik, Martin. 1960. Simulation of the Industry and The Firm. *The American Economic Review* 50 (5): 908–919.

Simonite, Tom. 2008. Should Every Computer Chip have a Cosmic Ray Detector? NewScientist. https://www.newscientist.com/blog/technology/2008/03/do-we-need-cosmic-ray-alerts-for.html.

Slayman, C. 2010. Soft Errors-Past History and Recent Discoveries. In 2010. *IEEE International Integrated Reliability Workshop Final Report* 25–30. IEEE Computer Society Press.

Smith, Reid G., and Randall Davis. 1981. Frameworks for Cooperation in Distributed Problem Solving. *IEEE Transactions on Systems, Man, and Cybernetics* 11 (1): 61–70. ISSN: 0018-9472. https://doi.org/10.1109/TSMC.1981.4308579.

Spier, Raymond E. 2012. Science and Engineering Ethics, Overview. In *Encyclopedia of Applied Ethics* (Second Edition), ed. by Ruth Chadwick, 14–31. Academic Press.

Spirtes, Peter, Clark Glymour, and Richard Scheines. 1993. *Causation, Prediction, and Search*. MIT Press.

Spivey, J.M. 2001. *The Z Notation: A Reference Manual*. Prentice Hall.

Steinle, Friedrich. 1997. Entering New Fields: Exploratory Uses of Experimentation. *Philosophy of Science* 64: S65–S74.

Steinle, Friedrich. 2002. Experiments in History and Philosophy of Science. *Perspectives on Science* 10 (4): 408–432.

Steup, Matthias., and Ernest Sosa (eds.). 2005. *Contemporary Debates in Epistemology*. Blackwell Publishing.

Stuewer, Roger H. 1985. Artificial Disintegration and the Cambridge-Vienna Controversy. In *Observation, Experiment, and Hypothesis in Modern Physical Science*, ed. by Peter Achinstein and Owen Hannaway, 239–307. MIT Press.

Sun, Ying, Gu Lianhong, Robert E. Dickinson, Richard J. Norby, Stephen G. Pallardy, and Forrest M. Hoffman. 2014. Impact of Mesophyll Diffusion on Estimated Global Land CO_2 Fertilization. *Proceedings of the National Academy of Sciences* 111 (44): 15774–15779.

Suppe, Frederick (ed.). 1977. *The Structure of Scientific Theories*. University of Illinois Press.

Tal, Eran. 2011. How Accurate is the Standard Second? *Philosophy of Science* 78 (5): 1082–1096.

Teichroew, Daniel, and John Francis Lubin. 1966. Computer Simulation—Discussion of the Technique and Comparison of Languages. *Communications of the ACM* 9 (10): 723–741.

Teller, Paul. 2013. The Concept of Measurement-Precision. *Synthese* 190 (2): 189–202.

Toffoli, Tommaso. 1984. CAM: A High-Performance Cellular-Automaton Machine. *Physica D: Nonlinear Phenomena* 10 (1–2): 195–204.

Tolk, Andreas. 2017. Code of Ethics. In *The Profession of Modeling and Simulation: Discipline, Ethics, Education, Vocation, Societies, and Economics*, ed. by Andreas Tolk and Tuncer Ören, 35–51. Wiley.

Trucano, Timothy G., L.P. Swiler, T. Igusa, W.L. Oberkampf, and M. Pilch. 2006. Calibration, Validation, and Sensitivity Analysis: What's What. *Reliability Engineering and System Safety* 91: 1331–1357.

Turner, Raymond. 2011. Specification. *Minds and Machines* 21 (2): 135–152.

U.S. Department of Health and Human Services. 2014. The Health Consequences of Smoking-50 Years of Progress: A Report of the Surgeon General—U.S. Department of Health and Human Services, Centers for Disease Control and Prevention, National Center for Chronic Disease Prevention and Health Promotion, Office on Smoking and Health. Atlanta, GA. https://www.ncbi. nlm.nih.gov/pubmed/24455788. Accessed 17 Oct 2016.

Vallverdú, Jordi. 2014. What are Simulations? An Epistemological Approach. *Procedia Technology* 13: 6–15.

van Helden, Albert. 1974. Saturn and his Anses. *Journal for the History of Astronomy* 5 (2): 105–121.

van der Poel, Ibo and Lambèr Royakkers 2011. Ethics, Technology, and Engineering. An Introduction. Wiley-Blackwell.

Vichniac, Gérard Y. 1984. Simulating Physics with Cellular Automata. *Physica D: Nonlinear Phenomena* 10: 96–116.

Von Neumann, John. 1945. First Draft of a Report on the EDAVAC. United States Army Ordnance Department University of Pennsylvania.

Waters, C.Kenneth. 2007. The Nature and Context of Exploratory Experimentation: An Introduction to Three Case Studies of Exploratory Research. *History and Philosophy of the Life Sciences* 29 (3): 275–284.

Weber, Marcel. 2005. *Philosophy of Experimental Biology*. Cambridge University Press.

Weirich, Paul. 2011. The Explanatory Power of Models and Simulations: A Philosophical Exploration. *Simulation & Gaming* 42 (2): 155–176.

Weisberg, Michael. 2013. *Simulation and Similarity*. Oxford University Press.

Wenham, C. Lawrence. 2012. *Signs That You're a Bad Programmer*. http://www.yacoset.com/ Home/signs-that-you-re-a-bad-programmer. Accessed 12 Apr 2016.

Wilhelmson, Robert, Lou Wicker, Matthew Gilmore, Glen Romine, Lee Cronce, Mark Straka, Donna Cox, et al. 2010. Visualization of an F3 Tornado: Storm Chaser Perspective. *Technical report. NCSA'S Advanced Visualization Laboratory*. http://avl.ncsa.illinois.edu/what-we-do/ services/media-downloadswatch?v=EgumU0Ns1YI. Accessed 10 Aug 2015.

Wilhelmson, Robert, Mathhew Gilmore, Louis Wicker, Glen Romine, Lee Cronce, and Mark Straka. 2005. Visualization of an F3 Tornado within a Simulated Supercell Thunderstorm. In *Proceeding SIGGRAPH '05 ACM SIGGRAPH 2005 Electronic Art and Animation Catalog*, 248–249. http:// avl.ncsa.illinois.edu/what-we-do/services/media-downloads Accessed 10 Aug 2015.

Williamson, Terence J. 2010. Predicting Building Performance: The Ethics of Computer Simulation. *Building Research & Information* 38 (4): 401–410.

Wilson, Curtis. 1993. Clairaut's Calculation of the Eighteenth-century Return of Halley's Comet. *Journal for the History of Astronomy* 24: 1–15.

Winsberg, Eric. 1999. Sanctioning Models: The Epistemology of Simulation. *Science in Context* 12: 275–292.

Winsberg, Eric. 2001. Simulations, Models, and Theories: Complex Physical Systems and Their Representations. *Philosophy of Science* 68 (3): S442–S454.

Winsberg, Eric. 2009. A Tale of Two Methods. *Synthese* 169: 575–592.

Winsberg, Eric. 2010. *Science in the Age of Computer Simulation*. University of Chicago Press.

Winsberg, Eric. 2015. Computer Simulations in Science. In *The Stanford Encyclopedia of Philosophy*, (Summer 2015 Edition), ed. by Edward N. Zalta. https://plato.stanford.edu/archives/ sum2018/entries/simulations-science/. Accessed 10 Sept 2015.

Wittgenstein, Ludwig. 1976. *Zettel*. Ed. by G.E. Anscombe and Georg Henrik Von Wright. University of California Press.

Wolfram, Stephen. 1984a. Preface. *Physica 10D*: vii–xii.

Wolfram, Stephen. 1984b. Universality and Complexity in Cellular Automata. *Physica 10D* 1–35.

Woodward, James. 2003. *Making Things Happen*. Oxford University Press.

Woolfson, Michael M., and Geoffrey J. Pert. 1999a. *An Introduction to Computer Simulations*. Oxford University Press.

Woolfson, Michael M., and Geoffrey J. Pert. 1999b. SATELLIT.FOR. *An Introduction to Computer Simulations*. Oxford University Press.

Wylie, Alison. 2002. *Thinking from Things: Essays in the Philosophy of Archeology*. University of California Press.

Yoshii, Yuzuru, Kentaro Motohara, Takashi Miyata, and Natsuko Mitani. 2009. The 1 m Telescope at the Atacama Observatory has Started Scientific Operation, Detecting the Hydrogen Emission Line from the Galactic Center in the Infrared Light. http://www.s.u-tokyo.ac.jp/en/press/2009/15.html. Accessed 12 Nov 2016.

Zenil, Hector. 2014. What Is Nature-Like Computation? A Behavioural Approach and a Notion of Programmability. *Philosophy&Technology* 27: 399–421.

Zhang, Zhongshi, Gilles Ramstein, Mathieu Schuster, Camille Li, Camille Contoux, and Qing Yan. 2014. Aridification of the Sahara Desert Caused by Tethys Sea Shrinkage During the Late Miocene. *Nature* 513 (7518): 401–404.

Zwitter, Andrej. 2014. Big Data Ethics. *Big Data & Society* 1 (2): 1–6.

Titles in This Series

Quantum Mechanics and Gravity
By Mendel Sachs

Quantum-Classical Correspondence
Dynamical Quantization and the Classical Limit
By A.O. Bolivar

Knowledge and the World: Challenges Beyond the Science Wars
Ed. by M. Carrier, J. Roggenhofer, G. Küppers and P. Blanchard

Quantum-Classical Analogies
By Daniela Dragoman and Mircea Dragoman

Quo Vadis Quantum Mechanics?
Ed. by Avshalom C. Elitzur, Shahar Dolev and Nancy Kolenda

Information and Its Role in Nature
By Juan G. Roederer

Extreme Events in Nature and Society
Ed. by Sergio Albeverio, Volker Jentsch and Holger Kantz

The Thermodynamic Machinery of Life
By Michal Kurzynski

Weak Links
The Universal Key to the Stability of Networks and Complex Systems
By Csermely Peter

The Emerging Physics of Consciousness
Ed. by Jack A. Tuszynski

Quantum Mechanics at the Crossroads
New Perspectives from History, Philosophy and Physics
Ed. by James Evans and Alan S. Thorndike

Mind, Matter and the Implicate Order
By Paavo T.I. Pylkkanen

Particle Metaphysics
A Critical Account of Subatomic Reality
By Brigitte Falkenburg

The Physical Basis of the Direction of Time
By H. Dieter Zeh

Asymmetry: The Foundation of Information
By Scott J. Muller

Decoherence
and the Quantum-To-Classical Transition
By Maximilian A. Schlosshauer

The Nonlinear Universe
Chaos, Emergence, Life
By Alwyn C. Scott

Quantum Superposition
Counterintuitive Consequences of Coherence, Entanglement, and Interference
By Mark P. Silverman

Symmetry Rules
How Science and Nature are Founded on Symmetry
By Joseph Rosen

Mind, Matter and Quantum Mechanics
By Henry P. Stapp

Entanglement, Information, and the Interpretation of Quantum Mechanics
By Gregg Jaeger

Relativity and the Nature of Spacetime
By Vesselin Petkov

The Biological Evolution of Religious Mind and Behavior
Ed. by Eckart Voland and Wulf Schiefenhövel

Homo Novus-A Human without Illusions
Ed. by Ulrich J. Frey, Charlotte Störmer and Kai P. Willführ

Brain-Computer Interfaces
Revolutionizing Human-Computer Interaction
Ed. by Bernhard Graimann, Brendan Allison and Gert Pfurtscheller

Extreme States of Matter
On Earth and in the Cosmos
By Vladimir E. Fortov

Searching for Extraterrestrial Intelligence
SETI Past, Present, and Future
Ed. by H. Paul Shuch

Essential Building Blocks of Human Nature
Ed. by Ulrich J. Frey, Charlotte Störmer and Kai P. Willführ

Mindful Universe
Quantum Mechanics and the Participating Observer
By Henry P. Stapp

Principles of Evolution
From the Planck Epoch to Complex Multicellular Life
Ed. by Hildegard Meyer-Ortmanns and Stefan Thurner

The Second Law of Economics
Energy, Entropy, and the Origins of Wealth
By Reiner Kümmel

States of Consciousness
Experimental Insights into Meditation, Waking, Sleep and Dreams
Ed. by Dean Cvetkovic and Irena Cosic

Elegance and Enigma
The Quantum Interviews
Ed. by Maximilian Schlosshauer

Humans on Earth
From Origins to Possible Futures
By Filipe Duarte Santos

Evolution 2.0
Implications of Darwinism in Philosophy and the Social and Natural Sciences
Ed. by Martin Brinkworth and Friedel Weinert

Probability in Physics
Ed. by Yemima Ben-Menahem and Meir Hemmo

Chips 2020
A Guide to the Future of Nanoelectronics
Ed. by Bernd Hoefflinger

From the Web to the Grid and Beyond
Computing Paradigms Driven by High-Energy Physics
Ed. by René Brun, Frederico Carminati and Giuliana Galli-Carminati

The Language Phenomenon
Human Communication from Milliseconds to Millennia
Ed. by P.-M. Binder and K. Smith

The Dual Nature of Life
Interplay of the Individual and the Genome
By Gennadiy Zhegunov

Natural Fabrications
Science, Emergence and Consciousness
By William Seager

Ultimate Horizons
Probing the Limits of the Universe
By Helmut Satz

Physics, Nature and Society
A Guide to Order and Complexity in Our World
By Joaquín Marro

Extraterrestrial Altruism
Evolution and Ethics in the Cosmos
Ed. by Douglas A. Vakoch

The Beginning and the End
The Meaning of Life in a Cosmological Perspective
By Clément Vidal

A Brief History of String Theory
From Dual Models to M-Theory
By Dean Rickles

Singularity Hypotheses
A Scientific and Philosophical Assessment
Ed. by Amnon H. Eden, James H. Moor, Johnny H. Søraker and Eric Steinhart

Why More Is Different
Philosophical Issues in Condensed Matter Physics and Complex Systems
Ed. by Brigitte Falkenburg and Margaret Morrison

Questioning the Foundations of Physics
Which of Our Fundamental Assumptions Are Wrong?
Ed. by Anthony Aguirre, Brendan Foster and Zeeya Merali

It From Bit or Bit From It?
On Physics and Information
Ed. by Anthony Aguirre, Brendan Foster and Zeeya Merali

How Should Humanity Steer the Future?
Ed. by Anthony Aguirre, Brendan Foster and Zeeya Merali

Trick or Truth?
The Mysterious Connection Between Physics and Mathematics
Ed. by Anthony Aguirre, Brendan Foster and Zeeya Merali

The Challenge of Chance
A Multidisciplinary Approach from Science and the Humanities
Ed. by Klaas Landsman, Ellen van Wolde

Quantum [Un]Speakables II
Half a Century of Bell's Theorem
Ed. by Reinhold Bertlmann, Anton Zeilinger

Energy, Complexity and Wealth Maximization
Ed. by Robert Ayres

Ancestors, Territoriality and Gods
A Natural History of Religion
By Ina Wunn, Davina Grojnowski

Space, Time and the Limits of Human Understanding
Ed. by Shyam Wuppuluri, Giancarlo Ghirardi

Information and Interaction
Eddington, Wheeler, and the Limits of Knowledge
Ed. by Ian T. Durham, Dean Rickles

The Technological Singularity
Managing the Journey
Ed. by V. Callaghan, J. Miller, R. Yampolskiy, S. Armstrong

How Can Physics Underlie the Mind?
Top-Down Causation in the Human Context
By George Ellis

The Unknown as an Engine for Science
An Essay on the Definite and the Indefinite
Hans J. Pirner

CHIPS 2020 Vol. 2
New Vistas in Nanoelectronics
Ed. by Bernd Hoefflinger

Life—As a Matter of Fat
Lipids in a Membrane Biophysics Perspective
Ole G. Mouritsen, Luis A. Bagatolli

The Quantum World
Philosophical Debates on Quantum Physics
Ed. by Bernard D'Espagnat, Hervé Zwirn

The Seneca Effect
Why Growth is Slow but Collapse is Rapid
By Ugo Bardi

Chemical Complexity
Self-Organization Processes in Molecular Systems
By Alexander S. Mikhailov, Gerhard Ertl

The Essential Tension
Competition, Cooperation and Multilevel Selection in Evolution
By Sonya Bahar

The Computability of the World
How Far Can Science Take Us?
By Bernd-Olaf Küppers

The Map and the Territory
Exploring the Foundations of Science, Thought and Reality
By Shyam Wuppuluri, Francisco A. Doria

Wandering Towards a Goal
How Can Mindless Mathematical Laws Give Rise to Aims and Intention?
Ed. by Anthony Aguirre, Brendan Foster and Zeeya Merali

Computer Simulations in Science and Engineering
Concepts—Practices—Perspectives
By Juan M. Durán

Stepping Stones to Synthetic Biology
By Sergio Carrà

Printed in the United States
By Bookmasters